Harriet Beecher Stowe

Flowers and fruit from the writings of Harriet Beecher Stowe

Harriet Beecher Stowe

Flowers and fruit from the writings of Harriet Beecher Stowe

ISBN/EAN: 9783337118273

Printed in Europe, USA, Canada, Australia, Japan

Cover: Foto ©berggeist007 / pixelio.de

More available books at **www.hansebooks.com**

Writings of Harriet Beecher Stowe.

UNCLE TOM'S CABIN. *Popular Illustrated Edition.* 12mo, $2.00.

THE SAME. *Illustrated Edition.* A new edition, from new plates, printed with red-line border. With an Introduction of more than 30 pages, and a bibliography of the various editions and languages in which the work has appeared, by Mr. GEORGE BULLEN, of the British Museum. Over 100 illustrations. 8vo, $3.50.

THE SAME. *Popular Edition.* With Introduction, and Portrait of " Uncle Tom." 12mo, $1.00.

DRED (sometimes called " Nina Gordon "). 12mo, $1.50.

THE MINISTER'S WOOING. 12mo, $1.50.

AGNES OF SORRENTO. 12mo, $1.50.

THE PEARL OF ORR'S ISLAND. 12mo, $1.50.

THE MAY-FLOWER, ETC. 12mo, $1.50.

OLDTOWN FOLKS. 12mo, $1.50.

SAM LAWSON'S FIRESIDE STORIES. New and enlarged Edition. Illustrated. 12mo, $1.50.

THE SAME. 16mo, paper covers, 50 cents.

MY WIFE AND I. New Edition. Illustrated. 12mo, $1.50.

WE AND OUR NEIGHBORS. New Edition. Illustrated. 12mo, $1.50.

POGANUC PEOPLE. New Edition. Illustrated 12mo, $1.50.

The above eleven 12mo volumes, uniform, in box, $16.50.

HOUSE AND HOME PAPERS. 16mo, $1.50.

LITTLE FOXES. 16mo, $1.50.

THE CHIMNEY-CORNER. 16mo, $1.50.

A DOG'S MISSION, ETC. New Edition. Illustrated. Small 4to, $1.25.

QUEER LITTLE PEOPLE. New Edition. Illustrated. Small 4to, $1.25.

LITTLE PUSSY WILLOW. New Edition. Illustrated. Small 4to, $1.25.

RELIGIOUS POEMS. Illustrated. 16mo, gilt edges, $1.50.

PALMETTO LEAVES. Sketches of Florida. Illustrated. 16mo, $1.50.

FLOWERS AND FRUIT. From Mrs. STOWE's Writings. 16mo, $1.00.

SCENES FROM MRS. STOWE'S WORKS. Paper, 15 cents.

HOUGHTON, MIFFLIN & CO., *Publishers,*

BOSTON.

FLOWERS AND FRUIT FROM THE WRITINGS OF HARRIET BEECHER STOWE

ARRANGED BY

ABBIE H. FAIRFIELD

BOSTON AND NEW YORK
HOUGHTON, MIFFLIN AND COMPANY
The Riverside Press, Cambridge
1888

CONTENTS.

FLOWERS AND FRUIT.

CHAPTER I.

THE INNER LIFE.

THE MINISTER'S WOOING.

Sympathy. When we feel a thing ourselves, we can see very quick the same in others.

Self-deception. When a finely constituted nature wishes to go into baseness, it has first to bribe itself. Evil is never embraced, undisguised, as evil, but under some fiction which the mind accepts, and with which it has the singular power of blinding itself in the face of daylight. The power of imposing on one's self is an essential preliminary to imposing on others. The man first argues himself down, and then he is ready to put the whole weight of his nature to deceiving others.

Soul-communion. Perhaps it is so, that souls once intimately related have ever after this a strange power of affecting each other, — a power that neither absence nor death can annul. How else can we interpret those mysterious hours in

which the power of departed love seems to overshadow us, making our souls vital with such long-
ings, with such wild throbbings, with such unutterable sighings, that a little more might burst
the mortal bond? Is it not deep calling unto
deep? the free soul singing outside the cage to
her mate beating against the bars within?

Soul-
absorption. The greatest moral effects are like those
of music, — not wrought out by sharp-
sided intellectual propositions, but melted in by
a divine fusion, by words that have mysterious,
indefinite fulness of meaning, made living by
sweet voices, which seem to be the out-throb-
bings of angelic hearts. So one verse in the
Bible read by a mother in some hour of tender
prayer has a significance deeper and higher than
the most elaborate of sermons, the most acute of
arguments.

Restric-
tions of
the body
on the
soul. Scarcely conscious, she lay in that dim,
clairvoyant state, when the half-sleep of
the outward senses permits a delicious
dewy clearness of the soul, that perfect
ethereal rest and freshness of the faculties, com-
parable only to what we imagine of the spiritual
state, — season of celestial enchantment, in which
the heavy weight " of all this unintelligible world "
drops off, and the soul, divinely charmed, nestles
like a wind-tossed bird in the protecting bosom
of the One All-perfect, All-beautiful. What visions then come to the inner eye have often no

words corresponding in mortal vocabularies. The poet, the artist, and the prophet in such hours become possessed of divine certainties which all their lives they struggle, with pencil or song or burning words, to make evident to their fellows. The world around wonders, but they are unsatisfied, because they have seen the glory and know how inadequate the copy.

Courage in the truth. Half the misery in the world comes of want of courage to speak and to hear the truth plainly and in a spirit of love.

Miss Prissy's motto. "Tho' I can't say I 'm lone either, because nobody need say that, so long as there 's folks to be done for."

Blessedness vs. happiness. We could not afford to have it always night, — and we must think that the broad, gay morning-light, when meadow-lark and robin and bobolink are singing in chorus with a thousand insects and the waving of a thousand breezes, is on the whole the most in accordance with the average wants of those who have a material life to live and material work to do. But then we reverence that clear-obscure of midnight, when everything is still and dewy ; — then sing the nightingales, which cannot be heard by day ; then shine the mysterious stars. So when all earthly voices are hushed in the soul, all earthly lights darkened, music and color float in from a higher sphere. . . . By what name shall we call this beautiful twilight, this night of the soul, so starry with heavenly mysteries ? *Not* happiness, but blessedness. They who have it walk among men as "sorrowful, yet alway rejoicing, as poor, yet making many rich, — as having nothing, and yet possessing all things."

Laws of prayer.

Is it not possible that He who made the world may have established laws for prayer as invariable as those for the sowing of seed and raising of grain ? Is it not as legitimate a subject of inquiry, when petitions are not answered, which of these laws has been neglected ?

Influence of the invisible.

No real artist or philosopher ever lived who has not at some hours risen to the height of utter self-abnegation for the glory of the invisible. There have been painters who would have been crucified to demonstrate the action of a muscle, — chemists who would gladly have melted themselves and all humanity in their crucible, if so a new discovery might arise out of its fumes. Even persons of mere artistic sensibility are at times raised by music, painting, or poetry, to a momentary trance of self-oblivion, in which they would offer their whole being before the shrine of an invisible loveliness. . . . But where theorists and

Woman feels ; man reasons.

philosophers tread with sublime assurance, woman often follows with bleeding footsteps; — women are always turning from the abstract to the individual, and feeling where the philosopher only thinks.

Love a sacrament. True love is a natural sacrament; and if ever a young man thanks God for having saved what is noble and manly in his soul, it is when he thinks of offering it to the woman he loves.

God's tests. "Mr. Scudder used to say that it took great affliction to bring his mind to that place," said Mrs. Katy. "He used to say that an old paper-maker told him once, that paper that was shaken only one way in the making would tear across the other, and the best paper had to be shaken every way; and so he said we could n't tell till we had been turned and shaken and tried every way, where we should tear."

Unconscious heart-thrusts. So we go, — so little knowing what we touch and what touches us as we talk! We drop out a common piece of news, — "Mr. So-and-so is dead," — "Miss Such-a-one is married," — "Such a ship has sailed," — and lo, on our right hand or our left, some heart has sunk under the news silently, — gone down in the great ocean of Fate, without even a bubble rising to tell its drowning pang. And this — God help us! — is what we call living!

Repression. It was not pride, nor sternness, but a sort of habitual shamefacedness that

kept far back in each soul those feelings which
are the most beautiful in their outcome; but
after a while the habit became so fixed a nature
that a caressing or affectionate expression could
not have passed the lips of one to another without a painful awkwardness.   Love was understood, once for all, to be the basis on which their
life was built.   Once for all, they loved each
other, and after that, the less said, the better.   It
had cost the woman's heart of Mrs. Marvin some
pangs, in the earlier part of her wedlock, to
accept of this *once for all* in place of those daily
outgushings which every woman desires should be
like God's loving-kindnesses, " new every morning ; " but hers, too, was a nature strongly inclin-
ing inward, and, after a few tremulous move-
ments, the needle of her soul settled, and her
life-lot was accepted, — not as what she would
like or could conceive, but as a reasonable and
good one.   Life was a picture painted in low,
cool tones, but in perfect keeping; and though
another and brighter style might have pleased
better, she did not quarrel with this.

Winged
and walk-
ing spirits.
There are in this world two kinds of
natures, — those that have wings, and
those that have feet, — the winged and
the walking spirits.   The walking are the logi-
cians ; the winged are the instinctive and poetic.
Natures that must always walk find many a bog,
many a thicket, many a tangled brake, which
God's happy little winged birds flit over by one

noiseless flight. Nay, when a man has toiled till his feet weigh too heavily with the mud of earth to enable him to walk another step, these little birds will often cleave the air in a right line towards the bosom of God, and show the way where he could never have found it.

Unity in prayer. The truly good are of one language in prayer. Whatever lines or angles of thought may separate them in other hours, *when they pray in extremity,* all good men pray alike. The Emperor Charles V. and Martin Luther, two great generals of opposite faiths, breathed out their dying struggles in the self-same words.

Sympathy through sorrow. As well might those on the hither side of mortality instruct the souls gone beyond the veil, as souls outside a great affliction guide those who are struggling in it. That is a mighty baptism, and only Christ can go down with us into those waters.

Agony of uncertainty. Against an uncertainty, who can brace the soul? We put all our forces of faith and prayer against it, and it goes down just as a buoy sinks in the water, and the next moment it is up again. The soul fatigues itself with efforts which come and go in waves; and when with laborious care it has adjusted all things in the light of hope, back flows the tide and sweeps all away. In such struggles life spends itself fast; an inward wound does not carry one deathward more surely than this worst wound of the soul. God has made us so mercifully that there is no *certainty*, however dreadful, to which life-forces do not in time adjust themselves; but to uncertainty there is no possible adjustment.

Candace's theology. " 'Cause, as we 's got to live in dis yer world, it 's quite clar de Lord must ha' fixed it so we *can*; an' ef tings was as some folks suppose, why, we could n't live, and dar would n't be no sense in anyting dat goes on."

Death in life. So we go, dear reader, — so long as we have a body and a soul. For worlds must mingle, — the great and the little, the solemn and the trivial, wreathing in and out, like the grotesque carvings on a gothic shrine ; only, did we know it rightly, nothing is trivial, since the human soul, with its awful shadow, makes all things sacred. Have not ribbons, cast-off flowers, soiled bits of gauze, trivial, trashy fragments of millinery, sometimes had an awful meaning, a deadly power, when they belonged to one who should wear them no more, and whose beautiful form, frail and crushed as they, is a hidden and vanished thing for all time ? For so sacred and individual is a human being, that, of all the million-peopled earth, no one form ever restores another. The mould of each mortal type is broken at the grave ; and never, never, though you look through all the faces on earth, shall the exact form you mourn ever meet your eyes again! You are living your daily life among trifles that one death-stroke may make relics. One false step, one luckless accident, an obstacle on the track of a train, the tangling of a cord in shifting a sail, and the penknife, the pen, the papers, the trivial articles of dress and clothing, which to-day you toss idly and jestingly from hand to hand, may become dread memorials of that awful tragedy whose deep abyss ever underlies our common life.

Memory. For one flower laid on the shrine which we keep in our hearts for the dead is worth more than any gift to our living selves.

Control of the thoughts. "How could you help it, *mignonne?* Can you stop your thinking?"
Mary said, after a moment's blush, —
"I can *try!*"

Minor modulations. Behind every scale of music, the gayest and cheeriest, the grandest, the most triumphant, lies its dark relative minor; the notes are the same, but the change of a semitone changes all to gloom; — all our gayest hours are tunes, that have a modulation into these dreary keys ever possible; at any moment the keynote may be struck.

The ideal and the practical. Nothing is more striking, in the light and shadow of the human drama, than to compare the inner life and thoughts of elevated and silent natures with the thoughts and plans which those by whom they are surrounded have of and for them. Little thought Mary of any of the speculations that busied the friendly head of Miss Prissy, or that lay in the provident forecastings of her prudent mother.

Perfect faith. "Indeed, I am afraid something must be wrong with me. I cannot have any fears, — I never could; I try sometimes, but the thought of God's goodness comes all around me, and I'm so happy before I think of it."

Undercurrent. All the little, mean work of our nature is generally done in a small dark closet just a little back of the subject we are talking about, on which subject we suppose ourselves of course to be thinking; — of course, we are thinking of it; how else could we talk about it?

The divine ideal. As to every leaf and every flower there is an ideal to which the growth of the plant is constantly urging, so is there an ideal to every human being, — a perfect form in which it might appear, were every defect removed, and every characteristic excellence stimulated to the highest point. Once in an age, God sends to some of us a friend who loves in us, not a false imagining, an unreal character, but, looking through all the rubbish of our imperfections, loves in us the divine ideal of our nature, —

loves, not the man that we are, but the angel that we may be.

Responsi-
bility.

To feel the immortality of a beloved soul hanging upon us, to feel that its only communications with Heaven must be through us, is the most solemn and touching thought that can pervade a mind.

Develop-
ing power
of love.

What makes the love of a great mind something fearful in its inception is, that it is often the unsealing of a hitherto undeveloped portion of a large and powerful being.

Unsus-
pected
influence.

It is said that if a grapevine be planted in the neighborhood of a well, its roots, running silently underground, wreathe themselves in a network round the cold, clear waters, and the vine's putting on outward greenness and unwonted clusters and fruit is all that tells where every root and fibre of its being has been silently stealing. So, those loves are most fatal, most absorbing, in which, with unheeded quietness, every thought and fibre of our life twines gradually around some human soul, to us the unsuspected wellspring of our being. Fearful it is, because so often the vine must be uprooted, and all its fibres wrenched away; but till the hour of discovery comes, how is it transfigured by a new and beautiful life!

Personal magnetism. Gradually an expression of intense interest and deep concern spread over the listeners ; it was the magnetism of a strong mind, which held them for a time under the shadow of his own awful sense of God's almighty justice.

It is said that a little child once described his appearance in the pulpit by saying, " I saw God there, and I was afraid."

Soul-growth. There is a ladder to heaven whose base God has placed in human affections, tender instincts, symbolic feelings, sacraments of love, through which the soul rises higher and higher, refining as she goes, till she outgrows the human, and changes, as she rises, into the image of the divine. At the very top of this ladder, at the threshold of Paradise, blazes dazzling and crystalline that celestial grade where the soul knows self no more, having learned, through a long experience of devotion, how blest it is to lose herself in that eternal Love and Beauty, of which all earthly fairness and grandeur are but the dim type, the distant shadow.

Discipline. It is said that gardeners sometimes, when they would bring a rose to richer flowering, deprive it, for a time, of light and moisture. Silent and dark it stands, dropping one faded leaf after another, and seeming to go down patiently to death. But when every leaf is dropped, and the plant stands stripped to the uttermost, a new life is even then working in the buds, from which shall spring a tender foliage, and a brighter wealth of flowers. So, often in celestial gardening, every leaf of earthly joy must drop, before a new and divine bloom visits the soul.

Idealizing power of love. In a refined and exalted nature, it is very seldom that the feeling of love, when once thoroughly aroused, bears any sort of relation to the reality of the object. It is commonly an enkindling of the whole power of the soul's love for whatever she considers highest and fairest; it is, in fact, the love of something divine and unearthly, which, by a sort of illusion, connects itself with a personality. Properly speaking, there is but one true, eternal object of all that the mind conceives, in this trance of its exaltation. Disenchantment must come, of course; and in a love which terminates in happy marriage, there is a tender and gracious process by which, without shock or violence, the ideal is gradually sunk in the real, which, though found faulty and earthly, is still ever tenderly remembered as it seemed under the morning light of that enchantment.

OLDTOWN FOLKS.

Moral earnestness. It is noticeable, in every battle of opinion, that honest, sincere, moral earnestness has a certain advantage over mere intellectual cleverness.

Struggling for higher things. Plato says that we all once had wings, and that they will tend to grow out in us, and that our burnings and aspirations for higher things are like the teething pangs of children. We are trying to cut our wings. Let us not despise these teething seasons. Though the wings do not become apparent, they may be starting under many a rough coat and on many a clumsy pair of shoulders.

Faith, not sight. "I often think," said Harry, listening for a moment, "that no one can pronounce on what this life has been to him until he has passed entirely through it, and turns around, and surveys it from the other world. I think then that we shall see everything in its true proportions; but, till then, we must walk by faith, not by sight, — faith that God loves us, faith that our Savior is always near us, and that all things are working together for good."

God's cordials. But certain it is that there is a very near way to God's heart, and so to the great heart of all comfort, that sometimes opens like a shaft of light between heaven and the soul, in hours when everything earthly falls away from us. A quaint old writer has said, "God keeps his choicest cordials for the time of our deepest faintings." And so it came to pass that, as this poor woman closed her eyes and prayed earnestly, there fell a strange clearness into her soul, which calmed every fear, and hushed the voice of every passion, and she lay for a season as if entranced. Words of Holy Writ, heard years ago in church readings, in the hours of unconscious girlhood, now seemed to come back, borne in with a loving power on her soul.

Silent companionship. The kind of silence which gives a sense of companionship.

Moral inheritance. Esther was one of those intense, silent, repressed women, that have been a frequent outgrowth of New England society. Moral traits, like physical ones, often intensify themselves in course of descent, so that the child of a long line of pious ancestry may sometimes suffer from too fine a moral fibre, and become a victim to a species of morbid *spiritual ideality.*

Esther looked to me, from the first, less like a warm, breathing, impulsive woman, less like ordinary flesh and blood, than some half-spiritual organization, every particle of which was a thought.

Holiness of age. Among all the loves that man has to woman, there is none so sacred and saint-like as that toward these dear, white-haired angels, who seem to form the connecting link between heaven and earth, who have lived to get the victory over every sin and every sorrow, and live perpetually on the banks of the dark river, in that bright, calm land of Beulah, where angels daily walk to and fro, and sounds of celestial music are heard across the water.

Such have no longer personal cares, or griefs, or sorrows. The tears of life have all been shed, and therefore they have hearts at leisure to attend to every one else. Even the sweet guileless childishness that comes on in this period has a sacred dignity; it is a seal of fitness for that heavenly kingdom, which whosoever shall not receive as a little child shall not enter therein.

Unity in conflict. Has there ever been a step in human progress that has not been taken against the prayers of some good soul, and been washed by tears sincerely and despondently shed? But, for all this, is there not a true unity of the faith in all good hearts? And when they have risen a little above the mists of earth, may not both sides — the conqueror and the conquered — agree that God hath given them the victory in advancing the cause of truth and goodness?

Growth from within. It has been the experience of my life that it is your quiet people who, above all other children of men, are set in their ways and intense in their opinions. Their very reserve and silence are a fortification behind which all their peculiarities grow and thrive at their leisure, without encountering those blows and shocks which materially modify more outspoken natures. It is owing to the peculiar power of quietness that one sometimes sees characters fashioning themselves in a manner the least to be expected from the circumstances and associates which surround them. As a fair white lily grows up out of the bed of meadow muck, and without note or comment rejects all in the soil that is alien from her being, and goes on fashioning her own silver cup side by side with weeds that are drawing coarser nutriment from the soil, so we often see a refined and gentle nature, by some singular internal force, unfolding itself by its own laws, and confirming itself in its own beliefs, as wholly different from all that surround it as is the lily from the ragweed. There are persons, in fact, who seem to grow almost wholly from within, and on whom the teachings, the doctrine, and the opinions of those around them, produce little or no impression.

Amusements. It may be set down, I think, as a general axiom, that people feel the need of amusements less and less, precisely in proportion as they have solid reasons for being happy.

Repression. Perhaps my readers may have turned over a great, flat stone some time in their rural rambles, and found under it little clovers, and tufts of grass pressed to earth, flat, white, and bloodless, but still growing, stretching, creeping towards the edges, where their plant-instinct tells them there is light and deliverance.

The kind of life that the little Tina led, under
the care of Miss Asphyxia, resembled that of
these poor clovers. It was all shut down and
repressed, but growing still.

Sympathy. I felt a cleaving of spirit to him that I
had never felt towards any human being before,
— a certainty that something had come to me in
him that I had always been wanting, — and I
was too glad for speech.

Soul-
relation.   Is it not true that, as we grow older,
the relationship of souls will make itself
felt ?

PALMETTO LEAVES.

Life
renewed.   No dreamland on earth can be more
unearthly in its beauty and glory than
the St. Johns in April. Tourists, for the most
part, see it only in winter, when half its gorgeous
forests stand bare of leaves, and go home, never
dreaming what it would be like in its resurrection robes. So do we, in our darkness, judge
the shores of the river of this mortal life up
which we sail, ofttimes disappointed and com-
plaining. We are seeing all things in winter,
and not as they will be when God shall wipe
away all tears, and bring about the new heavens
and new earth of which every spring is a symbol
and a prophecy. The flowers and leaves of last
year vanish for a season, but they come back
fresher and fairer than ever.

A lesson in faith. On either side, perched on a tall, dry, last year's coffee-bean-stalk, sit " papa " and " mamma," chattering and scolding, exhorting and coaxing. The little ones run from side to side, and say in plaintive squeaks, " I can't," " I dare n't," as plain as birds can say it. There, — now they spread their little wings; and oh, joy ! they find to their delight that they do not fall ; they exult in the possession of a new-born sense of existence. As we look at this pantomime, graver thoughts come over us. And we think how poor, timid, little souls moan, and hang back, and tremble, when the time comes to leave this nest of earth, and trust themselves to the free air of the world they were made for. As the little bird's moans and cries end in delight and rapture in finding himself in a new, glorious, free life ; so, just beyond the dark steps of death, will come a buoyant, exulting sense of new existence.

PEARL OF ORR'S ISLAND.

Discipline. The ship, built on one element, but designed to have its life in another, seemed an image of the soul, formed and fashioned with many a weary hammer-stroke in this life, but finding its true element only when it sails out into the ocean of eternity.

Heimweh. But there are souls sent into this world who seem to have always mysterious affinities

for the invisible and the unknown — who see the
face of everything beautiful through a thin veil
of mystery and sadness.   The Germans call this
yearning of spirit " homesickness " — the dim
remembrances of a spirit once affiliated to some
higher sphere, of whose lost brightness all things
fair are the vague reminders.

Limitation. But Miss Emily knew no more of the
deeper parts of her brother's nature than a little
bird that dips its beak into the sunny waters of
some spring knows of its depths of coldness and
shadow.

Learning   The fact was, as the reader may per-
to love.   ceive, that Miss Roxy had been thawed
into an unusual attachment for the little Mara,
and this affection was beginning to spread a
warming element through her whole being.   It
was as if a rough granite rock had suddenly
awakened to a passionate consciousness of the
beauty of some fluttering white anemone that
nestled in its cleft, and felt warm thrills running
through all its veins at every tender motion and
shadow.

Fitful   Such people are not very wholesome
persons.   companions for those who are sensitively
organized and predisposed to self-sacrificing love.
They keep the heart in a perpetual freeze and
thaw, which, like the American northern climate,
is so particularly fatal to plants of a delicate

habit. They could live through the hot summer and the cold winter, but they cannot endure the three or four months when it freezes one day and melts the next, — when all the buds are started out by a week of genial sunshine, and then frozen for a fortnight. These fitful persons are of all others most engrossing, because you are always sure in their good moods that they are just going to be angels, — an expectation which no number of disappointments seems finally to do away.

Love — a test.

Nothing so much shows what a human being is in moral advancement as the quality of his love.

LITTLE FOXES.

Altruistic faith.

The faults and mistakes of us poor human beings are as often perpetuated by despair as by any other one thing. Have we not all been burdened by a consciousness of faults that we were slow to correct because we felt discouraged? Have we not been sensible of a real help sometimes from the presence of a friend who thought well of us, believed in us, set our wisdom in the best light, and put our faults in the background?

Expression of love.

"Dispute your mother's hateful dogma, that love is to be taken for granted without daily proof between lovers; cry down latent caloric in the market; insist that the mere fact of being a wife is not enough, — that the words spoken once, years ago, are not enough, — that love needs new leaves every summer of life, as much as your elm-tree, and new branches to grow broader and wider, and new flowers at the root to cover the ground."

Latent caloric. I remember my school-day speculations over an old "Chemistry" I used to study as a text-book, which informed me that a substance called Caloric exists in all bodies. In some it exists in a latent state; it is there, but it affects neither the senses nor the thermometer. Certain causes develop it, when it raises the mercury and warms the hands. I remember the awe and wonder with which, even then, I reflected on the vast amount of blind, deaf, and dumb comfort which Nature had thus stowed away. How mysterious it seemed to me that poor families every winter should be shivering, freezing, and catching cold, when Nature had all this latent caloric locked up in her store-closet, — when it was all around them, in everything they touched and handled!

In the spiritual world there is an exact analogy to this. There is a great life-giving, warming power called Love, which exists in human hearts, dumb and unseen, but which has no real life, no warming power, till set free by expression.

Did you ever, in a raw, chilly day just before

a snowstorm, sit at work in a room that was judiciously warmed by an exact thermometer? You do not freeze, but you shiver; your fingers do not become numb with cold, but you have all the while an uneasy craving for more positive warmth. You look at the empty grate, walk mechanically towards it, and, suddenly awaking, shiver to see that there is nothing there. You long for a shawl or a cloak; you draw yourself within yourself; you consult the thermometer, and are vexed to find that there is nothing there to be complained of, — it is standing most provokingly at the exact temperature that all the good books and good doctors pronounce to be the proper thing, — the golden mean of health; and yet perversely you shiver, and feel as if the face of an open fire would be to you as the smile of an angel.

Such a life-long chill, such an habitual shiver, is the lot of many natures, which are not warm, when all ordinary rules tell them they ought to be warm, — whose life is cold and barren and meagre, — which never see the blaze of an open fire.

Regret. The bitterest tears shed over graves are for words left unsaid and deeds left undone. "She never knew how I loved her." "He never knew what he was to me." "I always meant to make more of our friendship." "I never knew what he was to me till he was gone." Such words are the poisoned arrows which cruel Death shoots backward at us from the door of the sepulchre.

How much more we might make of our family life, of our friendships, if every secret thought of love blossomed into a deed! We are not now speaking of personal caresses. These may or may not be the best language of affection. Many are endowed with a delicacy, a fastidiousness of physical organization, which shrinks away from too much of these, repelled and overpowered. But there are words and looks and little observances, thoughtfulnesses, watchful little attentions, which speak of love, which make it manifest, and there is scarce a family that might not be richer in heart-wealth for more of them.

HOUSE AND HOME PAPERS.

First principles of home-making. In this art of home-making I have set down in my mind certain first principles, like the axioms of Euclid, and the first is, —

No home is possible without love.

All business marriages and marriages of convenience, all mere culinary marriages and marriages of mere animal passion, make the creation of a true home impossible in the outset. Love is the jewelled foundation of this New Jerusalem descending from God out of heaven, and takes as many bright forms as the amethyst, topaz, and sapphire of that mysterious vision.

In this range of creative art all things are possible to him that loveth, but without love nothing is possible.

THE CHIMNEY CORNER.

Conversation. *Real* conversation presupposes intimate acquaintance. People must see each other often enough to wear off the rough bark and outside rind of commonplaces and conventionalities in which their real ideas are enwrapped, and give forth without reserve their innermost and best feelings.

Saintliness. What makes saintliness, in my view, as distinguished from ordinary goodness, is a certain quality of magnanimity and greatness of soul that brings life within the circle of the heroic. To be really great in little things, to be truly noble and heroic in the insipid details of every-day life, is a virtue so rare as to be worthy of canonization.

Teachings of suffering. There is a certain amount of suffering which must follow the rending of the great cords of life, suffering which is natural and inevitable : it cannot be argued down ; it cannot be stilled ; it can no more be soothed by any effort of faith and reason than the pain of a fractured limb, or the agony of fire on the living flesh. All that we can do is to brace ourselves to bear it, calling on God, as the martyrs did in the fire, and resigning ourselves to let it burn on. We must be willing to suffer, since God so wills. There are just so many waves to go over us, just so many arrows of stinging thought to be shot into our soul, just so many faintings and sinkings and revivings only to suffer again, belonging to and inherent in our portion of sorrow; and there is a work of healing that God has placed in the hands of Time alone.

Time heals all things at last; yet it depends much on us in our sufferings, whether Time shall send us forth healed, indeed, but maimed and crippled and callous, or whether, looking to the great Physician of sorrows, and coworking with him, we come forth stronger and fairer even for our wounds.

Help in sorrow. One soul redeemed will do more to lift the burden of sorrow than all the blandishments and diversions of art, all the alleviations of luxury, all the sympathy of friends.

THE MAYFLOWER.

Affinity of opposites. From that time a friendship commenced between the two which was a beautiful illustration of the affinities of opposites. It was like a friendship between morning and evening, — all freshness and sunshine on one side, and all gentleness and peace on the other.

Superiority. It is one mark of a superior mind to understand and be influenced by the superiority of others.

Sympathy. The same quickness which makes a mind buoyant in gladness often makes it gentlest and most sympathetic in sorrow.

God's sympathy. It is well for man that there is one Being who sees the suffering heart *as it is*, and not as it manifests itself through the repellences of outward infirmity, and who, perhaps, feels more for the stern and wayward than for those whose gentler feelings win for them human sympathy.

Influence. He had traced her, even as a hidden streamlet may be traced, by the freshness, the verdure of heart, which her deeds of kindness had left wherever she had passed.

Capacity of feeling. A very unnecessary and uncomfortable capacity of *feeling*, which, like a refined ear for music, is undesirable, because, in this world, one meets with discord ninety-nine times where one meets with harmony once.

Heart-wisdom *vs.* worldly wisdom. How very contrary is the obstinate estimate of the heart to the rational estimate of worldly wisdom! Are there not some who can remember when one word, one look, or even the withholding of a word, has

drawn their heart more to a person than all the
substantial favors in the world? By ordinary
acceptation, substantial kindness respects the
necessaries of animal existence, while those
wants which are peculiar to mind, and will exist
with it forever, by equally correct classification,
are designated as sentimental ones, the supply of
which, though it will excite more gratitude in
fact, ought not to in theory.

Living together. From that time I *lived* with her — and
there are some persons who can make
the word *live* signify much more than it commonly does — and she wrought on my character
all those miracles which benevolent genius can
work. She quieted my heart, directed my feelings, unfolded my mind, and educated me, not
harshly or by force, but as the blessed sunshine
educates the flower, into full and perfect life;
and when all that was mortal of her died to this
world, her words and deeds of unutterable love
shed a twilight around her memory that will fade
only in the brightness of heaven.

Minister-ing spirits. What then? May we look among the
band of ministering spirits for our own
departed ones? Whom would God be more
likely to send us? Have we in heaven a friend
who knew us to the heart's core? a friend to
whom we have unfolded our soul in its most
secret recesses? to whom we have confessed our
weaknesses and deplored our griefs? If we are

to have a ministering spirit, who better adapted? Have we not memories which correspond to such a belief? When our soul has been cast down, has never an invisible voice whispered, "There is lifting up"? Have not gales and breezes of sweet and healing thought been wafted over us, as if an angel had shaken from his wings the odors of paradise? Many a one, we are confident, can remember such things, — and whence come they? Why do the children of the pious mother, whose grave has grown green and smooth with years, seem often to walk through perils and dangers fearful and imminent as the crossing of Mohammed's fiery gulf on the edge of a drawn sword, yet walk unhurt? Ah! could we see that attendant form, that face where the angel conceals not the mother, our question would be answered.

Influence of a mother's prayer. Something there is in the voice of real prayer that thrills a child's heart, even before he understands it; the holy tones are a kind of heavenly music, and far-off in distant years, the callous and worldly man often thrills to his heart's core, when some turn of life recalls to him his mother's prayer.

PINK AND WHITE TYRANNY.

Taught by suffering. It sometimes seems to take a stab, a thrust, a wound, to open in some hearts the capacity of deep feeling and deep thought. There are things taught by suffering that can be taught in no other way. By suffering sometimes is wrought out in a person the power of loving and of appreciating love. During the first year, Lillie had often seemed to herself in a sort of wild, chaotic state. The coming in of a strange, new, spiritual life was something so inexplicable to her that it agitated and distressed her; and sometimes, when she appeared more petulant and fretful than usual, it was only the stir and vibration on her weak nerves of new feelings, which she wanted the power to express. These emotions at first were painful to her. She felt weak, miserable, and good-for-nothing. It seemed to her that her whole life had been a wretched cheat, and that she had ill repaid the devotion of her husband. At first these thoughts only made her bitter and angry; and she contended against them. But, as she sank from day to day, and grew weaker and weaker, she grew more gentle; and a better spirit seemed to enter into her.

The object of life. "The great object of life is not happiness; and when we have lost our own personal happiness, we have not lost all that life is worth living for. No, John, the very best of life often lies beyond that. When we have learned to let ourselves go, then we may find that there is a better, a nobler, and a truer life for us." . . . "If we contend with, and fly from our duties, simply because they gall us and bur-

den us, we go against everything; but if we take
them up bravely, then everything goes with us.
God and good angels and good men and all good
influences are working with us when we are
working for the right. And in this way, John,
you may come to happiness; or, if you do not
come to personal happiness, you may come to
something higher and better. You know that
you think it nobler to be an honest man than a
rich man; and I am sure that you will think
it better to be a good man than to be a happy
one."

Self-
ignorance. It is astonishing how blindly people
sometimes go on as to the character of
their own conduct, till suddenly, like a torch in a
dark place, the light of another person's opinion
is thrown in upon them, and they begin to judge
themselves under the quickening influence of
another person's moral magnetism. Then, indeed, it often happens that the graves give up
their dead, and that there is a sort of interior
resurrection and judgment.

Sympathy. When we are feeling with the nerves of
some one else, we notice every roughness and inconvenience.

Clairvoy-
ance. A terrible sort of clairvoyance that
seems to beset very sincere people, and
makes them sensitive to the presence of anything
unreal or untrue.

Unacknow-
ledged
motives.
No, she did not say it. It would be well for us all if we *did* put into words, plain and explicit, many instinctive resolves and purposes that arise in our hearts, and which, for want of being so expressed, influence us undetected and unchallenged. If we would say out boldly, " I don't care for right or wrong, or good or evil, or anybody's rights or anybody's happiness, or the general good, or God himself, — all I care for, or feel the least interest in, is to have a good time myself, and I mean to do it, come what may," — we should be only expressing a feeling which often lies in the dark backroom of the human heart; and saying it might alarm us from the drugged sleep of life. It might rouse us to shake off the slow, creeping paralysis of selfishness and sin before it is forever too late.

BETTY'S BRIGHT IDEA.

Aspiration. That noble discontent that rises to aspiration for higher things.

DEACON PITKIN'S FARM.

The lesson
of faith.
" Well, daughter," said the deacon, " it 's a pity we should go through all we do in this world and not learn anything by it. I hope the Lord has taught me not to worry, but

just do my best, and leave myself and everything else in his hands. We can't help ourselves, — we can't make one hair white or black. Why should we wear our lives out fretting? If I'd a known *that* years ago, it would a been better for us all."

"All for the best." "She's allers sayin' things is for the best, maybe she'll come to think so about this, — folks gen'ally does when they can't help themselves."

Sympathy. Eyes that have never wept cannot comprehend sorrow.

Trust. "Leave it!"

These were words often in that woman's mouth, and they expressed that habit of her life which made her victorious over all troubles, that habit of trust in the Infinite Will that actually could and did *leave* every accomplished event in his hand without murmur and without conflict.

AGNES OF SORRENTO.

Power of sympathy. Such is the wonderful power of human sympathy that the discovery even of the existence of a soul capable of understanding our inner life often operates as a perfect charm; every thought and feeling and aspiration carries with it a new value, from the interwoven con·

sciousness that attends it, of the worth it would bear to that other mind; so that, while that person lives, our existence is doubled in value, even though oceans divide us.

Difficulty of inspiring others. But he soon discovered, what every earnest soul learns who has been baptized into a sense of things invisible, how utterly powerless and inert any mortal man is to inspire others with his own insights and convictions. With bitter discouragement and chagrin, he saw that the spiritual man must forever lift the dead weight of all the indolence and indifference and animal sensuality that surround him, — that the curse of Cassandra is upon him, forever to burn and writhe under awful visions of truths which no one around him will regard.

Good wherever we seek it. As a bee can extract pure honey from the blossoms of some plants whose leaves are poisonous, so some souls can nourish themselves only with the holier and more ethereal parts of popular belief.

Naïveté. "Blessed are the flowers of God that grow in cool solitudes, and have never been profaned by the hot sun and dust of this world."

Sorrow a preparation for love. Never does love strike so deep and immediate a root as in a sorrowful and desolated nature; there it has nothing to dispute the soil, and soon fills it with its interlacing fibre.

Sunshine of the heart. "He is happy, like the birds," said Agnes, "because he flies near heaven."

Dreams. Dreams are the hushing of the bodily senses, that the eyes of the spirit may open.

Lost innocence irrecoverable. When a man has once lost that unconscious soul-purity which exists in a mind unscathed by the fires of passion, no after-tears can weep it back again. No penance, no prayer, no anguish of remorse, can give back the simplicity of a soul that has never been stained,

The strongest passions. No passions are deeper in their hold, more pervading and more vital to the whole human being, than those that make their first entrance through the higher nature, and, beginning with a religious and poetic ideality, gradually work their way through the whole fabric of the human existence. From grosser passions, whose roots lie in the senses, there is always a refuge in man's loftier nature. He can cast them aside with contempt, and leave them as one whose lower story is flooded can remove to a higher loft, and live serenely with a purer air and wider prospect. But to love that is born of ideality, of intellectual sympathy, of harmonies of the spiritual and immortal nature, of the very poetry and purity of the soul, if it be placed where reason and religion forbid its exercise and expression, what refuge but the grave, —

what hope but that wide eternity where all human barriers fall, all human relations end, and love ceases to be a crime.

Agony in the voice. It is singular how the dumb, imprisoned soul, locked within the walls of the body, sometimes gives such a piercing power to the tones of the voice during the access of a great agony. The effect is entirely involuntary and often against the most strenuous opposition of the will ; but one sometimes hears another reading or repeating words with an intense vitality, a living force, which tells of some inward anguish or conflict of which the language itself gives no expression.

A sympathetic God. The great Hearer of Prayer regards each heart in its own scope of vision, and helps not less the mistaken than the enlightened distress. And for that matter, who is enlightened? who carries to God's throne a trouble or a temptation in which there is *not* somewhere a misconception or a mistake?

Transient uplifting. We hold it better to have even transient upliftings of the nobler and more devout element of man's nature than never to have any at all, and that he who goes on in worldly and sordid courses, without ever a spark of religious enthusiasm or a throb of aspiration, is less of a man than he who sometimes soars heavenward, though his wings be weak and he fall again.

Coinci- When a man has a sensitive or sore
dence. spot in his heart, from the pain of
which he would gladly flee to the ends of the
earth, it is marvellous what coincidences of events
will be found to press upon it wherever he
may go.

Silence of They both sat awhile in that kind of
deep emo- quietude which often falls between two
tion. who have stirred some deep fountain of
emotion.

Innocence. There is something pleading and pitiful
in the simplicity of perfect ignorance, — a rare
and delicate beauty in its freshness, like the
morning-glory cup, which, once withered by the
heat, no second morning can restore.

World "This is such a beautiful world," said
conflicts. Agnes, "who would think it would be
such a hard one to live in? — such battles and
conflicts as people have here!"

Nervous As one looking through a prism sees a
sensibility. fine bordering of rainbow on every object, so he beheld a glorified world. His former
self seemed to him something forever past and
gone. He looked at himself as at another person, who had sinned and suffered, and was now
resting in beatified repose; and he fondly thought
all this was firm reality, and believed that he
was now proof against all earthly impressions,

able to hear and to judge with the dispassionate calmness of a disembodied spirit. He did not know that this high-strung calmness, this fine clearness, were only the most intense forms of nervous sensibility, and as vividly susceptible to every mortal impression as is the vitalized chemical plate to the least action of the sun's rays.

UNCLE TOM'S CABIN.

Sorrow an educator.

Any mind that is capable of a *real sorrow* is capable of good.

Individuality.

Now, the reflections of two men, sitting side by side, are a curious thing, — seated on the same seat, having the same eyes, ears, hands, and organs of all sorts, and having pass before their eyes the same objects, — it is wonderful what a variety we shall find in these same reflections.

Inspiration.

By what strange law of mind is it that an idea, long overlooked, and trodden under foot as a useless stone, suddenly sparkles out in new light, as a discovered diamond.

Power of mind over body.

Sublime is the dominion of the mind over the body, that, for a time, can make flesh and nerve impregnable, and string the sinews like steel, so that the weak become so mighty.

True heroism. Have not many of us, in the weary way of life, felt, in some hours, how far easier it were to die than to live ?

The martyr, when faced even by a death of bodily anguish and horror, finds in the very terror of his doom a strong stimulant and tonic. There is a vivid excitement, a thrill and fervor, which may carry through any crisis of suffering that is the birth-hour of eternal glory and rest.

But to live, — to wear on, day after day, of mean, bitter, low, harassing servitude, every nerve dampened and depressed, every power of feeling gradually smothered, — this long and wasting heart martyrdom, this slow, daily bleeding away of the inward life, drop by drop, hour after hour, — this is the true searching test of what there may be in man or woman.

Moral atmosphere. An atmosphere of sympathetic influence encircles every human being; and the man or woman who *feels* strongly, healthily, and justly, on the great interests of humanity, is a constant benefactor to the human race.

Self-sacrifice. There are in this world blessed souls, whose sorrows all spring up into joys for others; whose earthly hopes, laid in the grave with many tears, are the seed from which spring healing flowers and balm for the desolate and the distressed.

Strength of despair. When a heavy weight presses the soul to the lowest level at which endurance

is possible, there is an instant and desperate
effort of every physical and moral nerve to throw
off the weight; and hence the heaviest anguish
often precedes a return tide of joy and courage.

Self-forget-
fulness.
"Thee uses thyself only to learn how to
love thy neighbor, Ruth," said Simeon,
looking with a beaming face on Ruth.

Natural
religious
sensibility.
He had one of those natures which
could better and more clearly conceive
of religious things from its own percep-
tions and instincts than many a matter-of-fact and
practical Christian. The gift to appreciate and
the sense to feel the finer shades and relations of
moral things often seems an attribute of those
whose whole life shows a careless disregard of
them. Hence, Moore, Byron, Goethe, often speak
words more wisely descriptive of the true religious sentiment, than another man whose whole
life is governed by it. In such minds, disregard
of religion is a more fearful treason, — a more
deadly sin.

Supersti-
tion.
No one is so thoroughly superstitious as
the godless man. The Christian is com-
posed by the belief of a wise, all-ruling Father,
whose presence fills the void unknown with light
and order; but to the man who has dethroned
God, the spirit-land is, indeed, in the words of
the Hebrew poet, "a land of darkness and the
shadow of death," without any order, where the

light is as darkness. Life and death to him are haunted grounds, filled with goblin forms of vague and shadowy dread.

The human soul. After all, let a man take what pains he may to hush it down, a human soul is an awful ghostly, unquiet possession for a bad man to have. Who knows the metes and bounds of it? Who knows all its awful perhapses, — those shudderings and tremblings which it can no more live down than it can outlive its own eternity! What a fool is he who locks his door to keep out spirits, who has in his own bosom a spirit he dares not meet alone, — whose voice, smothered far down, and piled over with mountains of earthliness, is yet like the forewarning trumpet of doom!

DRED.

Practical and ideal. The divine part of man is often shame-faced and self-distrustful, ill at home in this world, and standing in awe of nothing so much as what is called common sense; and yet common sense very often, by its own keenness, is able to see that these unavailable currencies of another's mind are of more worth, if the world only knew it, than the ready coin of its own; and so the practical and the ideal nature are drawn together.

Inexplicable preferences. Sensitive people never like the fatigue of justifying their instincts. Nothing, in fact, is less capable of being justified by technical reasons than those fine insights into character whereupon affection is built. We have all had experience of preferences which would not follow the most exactly ascertained catalogue of virtues, and would be made captive where there was very little to be said in justification of the captivity.

Congeniality of opposites. " Why, surely," said Anne, " one wants one's friends to be congenial, I should think."

" So we do ; and there is nothing in the world so congenial as differences. To be sure, the differences must be harmonious. In music, now, for instance, one does n't want a repetition of the same notes, but differing notes that chord. Nay, even discords are indispensable to complete harmony. Now, Nina has just that difference from me which chords with me ; and all our little quarrels — for we have had a good many, and I dare say shall have more — are only a sort of chromatic passages, — discords of the seventh, leading into harmony. My life is inward, theorizing, self - absorbed. I am hypochondriac, often morbid. The vivacity and acuteness of her outer life makes her just what I need. She wakens, she rouses, and keeps me in play ; and her quick instincts are often more than a match for my reason."

Proof of heaven. " How do you know there is any heaven, anyhow ? "

" Know it ? " said Milly, her eyes kindling, and striking her staff on the ground, " Know it ? I know it by de *hankering arter it* I got in here ; " giving her broad chest a blow which made it resound like a barrel. " De Lord knowed what he was 'bout when he made us. When he made babies rootin' 'round, wid der poor little mouths open, he made milk, and de mammies for 'em too. Chile, we 's nothing but great babies, dat ain't got our eyes open, — rootin' 'round an' 'round ; but de Father 'll feed us yet — He will so."

Power of song. As oil will find its way into crevices where water cannot penetrate, so song will find its way where speech can no longer enter.

Night resolutions. What we have thought and said under the august presence of witnessing stars, or beneath the holy shadows of moonlight, seems with the dry, hot heat of next day's sun to take wings, and rise to heaven with the night's clear drops. If all the prayers and good resolutions which are laid down on sleeping pillows could be found there on awaking, the world would be better than it is.

Transition periods. There are times in life when the soul, like a half-grown climbing vine, hangs hovering tremulously, stretching out its tendrils for something to ascend by. Such are generally the great transition periods of life, when we are passing from the ideas and conditions of one stage of existence to those of another. Such times are most favorable for the presentation of the higher truths of religion.

Connection with the spirit world.
This life may truly be called a haunted house, built as it is on the very confines of the land of darkness and the shadow of death. A thousand living fibres connect us with the unknown and unseen state ; and the strongest hearts, which never stand still for any mortal terror, have sometimes hushed their very beating at a breath of a whisper from within the veil. Perhaps the most resolute unbeliever in spiritual things has hours of which he would be ashamed to tell, when he, too, yields to the powers of those awful affinities which bind us to that unknown realm.

Suffering in silence.
It is the last triumph of affection and magnanimity, when a loving heart can respect the suffering silence of its beloved, and allow that lonely liberty in which only some natures can find comfort.

Joy in endurance.
And, as he sang and prayed, that strange joy arose within him, which, like the sweetness of night flowers, is born of darkness and tribulation. The soul has in it

somewhat of the divine, in that it can have joy
in endurance beyond the joy of indulgence.

They mistake who suppose that the highest
happiness lies in wishes accomplished — in prosperity, wealth, favor, and success. There has
been a joy in dungeons and on racks passing the
joy of harvest. A joy strange and solemn,
mysterious even to its possessor. A white stone
dropped from that signet ring, peace, which a
dying Saviour took from his own bosom, and
bequeathed to those who endure the cross, de-
spising the shame.

SUNNY MEMORIES OF FOREIGN LANDS.

Inward
peace.
How natural it is to say of some place
sheltered, simple, cool, and retired, here
one might find peace, as if peace came from
without, and not from within. In the shadiest
and stillest places may be the most turbulent
hearts, and there are hearts which, through the
busiest scenes, carry with them unchanging peace.

Grace in
affliction.
I have read of Alpine flowers leaning
their cheeks on the snow. I wonder if
any flowers grow near enough to that snow to
touch it. I mean to go and see. So I went;
there, sure enough, my little fringed purple bell,
to which I had give the name of "suspirium,"
was growing, not only close to the snow but in it.

Thus God's grace, shining steadily on the

waste places of the human heart, brings up heavenward sighings and aspirations, which pierce through the cold snows of affliction, and tell that there is yet life beneath.

God as an artist. I was glad to walk on alone : for the scenery was so wonderful that human sympathy and communion seemed to be out of the question.   The effect of such scenery to our generally sleeping and drowsy souls, bound with a double chain of earthliness and sin, is like the electric touch of the angel on Peter, bound and sleeping.   They make us realize that we were not only made to commune with God, but also what a God He is with whom we may commune. We talk of poetry, we talk of painting, we go to the ends of the earth to see the artists and great men of this world; but what a poet, what an artist, is God!   Truly said Michel Angelo, " The true painting is only a copy of the divine perfections — a shadow of his pencil."

Soul-striving. The human soul seems to me an imprisoned essence, striving after somewhat divine.   There is strength in it, as of suffocated flame, finding vent now through poetry, now in painting, now in music, sculpture, or architecture ; various are the crevices and fissures, but the flame is one.

Shadow. What a curious kind of thing shadow is, — that invisible veil, falling so evenly and so

lightly over all things, bringing with it such
thoughts of calmness and rest. I wonder the old
Greeks did not build temples to Shadow, and call
her the sister of Thought and Peace. The
Hebrew writers speak of the "overshadowing
of the Almighty;" they call his protection "the
shadow of a great rock in a weary land." Even
as the shadow of Mont Blanc falls like a Sabbath
across this valley, so falls the sense of his presence across our weary life-road.

Heimweh. Why? why this veil of dim and inde-
finable anguish at sight of whatever is most fair,
at hearing whatever is most lovely? Is it the
exiled spirit, yearning for its own? Is it the
captive, to whom the ray of heaven's own glory
comes through the crevice of his dungeon wall?

Seeing and It is not enough to open one's eyes on
feeling. scenes; one must be able to be "en
rapport" with them. Just so in the spiritual
world, we sometimes *see* great truths, — see that
God is beautiful and surpassingly lovely; but at
other times we *feel* both nature and God, and O,
how different *seeing* and *feeling* !

POGANUC PEOPLE.'

Longing There are hard, sinful, unlovely souls,
for love in
the un- who yet long to be loved, who sigh in
lovely. their dark prison for that tenderness,

that devotion, of which they are consciously unworthy. Love might redeem them; but who can love them? There is a fable of a prince, doomed by a cruel enchanter to wear a loathsome, bestial form, till some fair woman should redeem him by the transforming kiss of love. The fable is a parable of the experience of many a lost human soul. . . .

Who can read the awful mysteries of a single soul? We see human beings, hard, harsh, earthly, and apparently without an aspiration for anything high and holy; but let us never say that there is not far down in the depths of any soul a smothered aspiration, a dumb, repressed desire to be something higher and purer, to attain the perfectness to which God calls it.

LITTLE PUSSY WILLOW.

Seeing the bright side.

"She shall be called little Pussy Willow, and I shall give her the gift of *always seeing the bright side of everything.* That gift will be more to her than beauty or riches or honors. It is not so much matter what color one's eyes are as what one sees with them. There is a bright side to everything, if people only knew it, and the best eyes are those which are always able to see this bright side."

A DOG'S MISSION.

Reaction of harshness. A conscientious person should beware of getting into a passion, for every sharp word one speaks comes back and lodges like a sliver in one's own heart; and such slivers hurt us worse than they ever can any one else.

Man's childish impatience. Ah, the child is father of the man! when he gets older he will have the great toys of which these are emblems ; he will believe in what he sees and touches, — in house, land, railroad stock, — he will believe in these earnestly and really, and in his eternal manhood nominally and partially. And when his father's messengers meet him, and face him about, and take him off his darling pursuits, and sweep his big ships into the fire, and crush his full-grown cars, then the grown man will complain and murmur, and wonder as the little man does now. The Father wants the future, the Child the present, all through life, till death makes the child a man.

MY WIFE AND I.

Discipline of patience. The moral discipline of bearing with evil patiently is a great deal better and more ennobling than the most vigorous assertion of one's personal rights.

Ennobling power of sorrow. When we look at the apparent recklessness with which great sorrows seem to be distributed among the children of the earth, there is no way to keep our faith in a Fatherly love, except to recognize how invariably the sorrows that spring from love are a means of enlarging and dignifying a human being. Nothing great or good comes without birth-pangs, and in just the proportion that natures grow more noble their capacities of suffering increase.

Line between right and wrong. The line between right and wrong seems always so indefinite, like the line between any two colors of the prism ; it is hard to say just where one ends and another begins.

Doubt. " Doubt is very well as a sort of constitutional crisis in the beginning of one's life ; but if it runs on and gets to be chronic it breaks a fellow up, and makes him morally spindling and sickly. Men that *do* anything in the world must be men of strong convictions ; it won't do to go through life like a hen, craw-crawing and lifting up one foot, not knowing where to set it down next."

Friends. " I don't think," said she, " you should say ' *make* ' friends, — friends are *discovered*, rather than made. There are people who are in their own nature friends, only they don't know each other ; but certain things, like poetry, music, and painting, are like the free-masons' signs, — they reveal the initiated to each other."

WE AND OUR NEIGHBORS.

Forgiveness of friends. "Yes," said Harry, "forgiveness of enemies used to be the *ultima thule* of virtue; but I rather think it will have to be forgiveness of friends. I call the man a perfect Christian that can always forgive his friends."

Altruism. Do not our failures and mistakes often come from discouragement? Does not every human being need a believing second self, whose support and approbation shall reinforce one's failing courage? The saddest hours of life are when we doubt ourselves. To sensitive, excitable people, who expend nervous energy freely, must come many such low tides. "Am I really a miserable failure, — a poor, good-for-nothing, abortive attempt?" In such crises we need another self to restore our equilibrium.

Reproach. The agony of his self-reproach and despair had been doubled by the reproaches and expostulations of many of his own family friends, who poured upon bare nerves the nitric acid of reproach.

Help from work. Something definite to do is, in some crises, a far better medicine for a sick soul than any amount of meditation and prayer. One step fairly taken in a right direction goes farther than any amount of agonized back-looking.

Praise and blame. Praise is sunshine; it warms, it inspires, it promotes growth: blame and rebuke are rain and hail; they beat down and bedraggle, even though they may at times be necessary.

God working through man. The invisible Christ must be made known through human eyes; He must speak though a voice of earthly love, and a human hand inspired by his spirit must be reached forth to save.

Inner life. The external life is positive, visible, definable; easily made the subject of conversation. The inner life is shy, retiring, most difficult to be expressed in words, often inexplicable, even to the subject of it, yet no less a positive reality than the outward.

RELIGIOUS POEMS.

Peace through suffering. For not alone in those old Eastern regions
Are Christ's beloved ones tried by cross and chain;
In many a house are his elect ones hidden,
His martyrs suffering in their patient pain.
The rack, the cross, life's weary wrench of woe,

The world sees not, as slow, from day to day,
In calm, unspoken patience, sadly still,
The loving spirit bleeds itself away;
But there are hours, when from the heavens un-
    folding
Come down the angels with the glad release,
And we look upward, to behold in glory
Our suffering loved ones borne away to peace.

The spirit within. As some rare perfume in a vase of clay
    Pervades it with a fragrance not its
        own,
So, when Thou dwellest in a mortal soul,
All heaven's own sweetness seems around it
    thrown.

The calm of God's love. When winds are raging o'er the upper
    ocean,
    And billows wild contend with angry
        roar,
'T is said, far down beneath the wild commotion,
That peaceful stillness reigneth evermore.
Far, far beneath, the noise of tempest dieth,
And silver waves chime ever peacefully;
And no rude storm, how fierce soe'er he flieth,
Disturbs the Sabbath of that deeper sea.
So to the soul that knows thy love, O Purest,
There is a temple peaceful evermore!
And all the babble of life's angry voices
Die in hushed stillness at its sacred door.

God's com-fort. Think not, when the wailing winds of
    autumn

Drive the shivering leaflets from the trees, —
Think not all is over : spring returneth ;
Buds and leaves and blossoms thou shalt see.
Think not, when thy heart is waste and dreary,
When thy cherished hopes lie chill and sere, —
Think not all is over: God still loveth ;
He will wipe away thy every tear.

CHAPTER II.

HUMAN NATURE.

THE MINISTER'S WOOING.

Ignorant selfishness. He was one of that class of people who, of a freezing day, will plant themselves directly between you and the fire, and then stand and argue to prove that selfishness is the root of all moral evil. Simeon said he always had thought so; and his neighbors sometimes supposed that nobody could enjoy better experimental advantages for understanding the subject. He was one of those men who suppose themselves submissive to the divine will, to the uttermost extent demanded by the extreme theology of that day, simply because they have no nerves to feel, no imagination to conceive, what endless happiness or suffering is, and who deal therefore with the great question of the salvation or damnation of myriads as a problem of theological algebra, to be worked out by their inevitable x, y, z.

Sensitiveness to blame. A generous, upright nature is always more sensitive to blame than another, — sensitive in proportion to the amount of its reverence for good.

Depression after exaltation. It is a hard condition of our existence that every exaltation must have its depression. God will not let us have heaven here below, but only such glimpses and faint showings as parents sometimes give to children, when they show them beforehand the jewelry and pictures and stores of rare and curious treasures which they hold for the possession of their riper years. So it very often happens that the man who has gone to bed an angel, feeling as if all sin were forever vanquished, and he himself immutably grounded in love, may wake the next morning with a sick-headache, and, if he be not careful, may scold about his breakfast like a miserable sinner.

French nature. True Frenchwoman as she was, always in one rainbow shimmer of fancy and feeling, like one of those cloud-spotted April days, which give you flowers and rain, sun and shadow, and snatches of bird-singing, all at once.

Simple honesty vs. worldliness. He is one of those great, honest fellows, without the smallest notion of the world we live in, who think, in dealing with men, that you must go to work and prove the right or the wrong of a matter; just as if anybody cared for that! Supposing he is right, — which appears very probable to me, — what is he going to do about it? No moral argument, since the world began, ever prevailed over twenty-five per cent. profit.

Duty *vs.* expediency.

"Madam," said the doctor, "I'd sooner my system should be sunk in the sea than that it should be a millstone round my neck to keep me from my duty. Let God take care of my theology; I must do my duty."

Joy of living.

There are some people so evidently broadly and heartily of this world that their coming into a room always materializes the conversation. We wish to be understood that we mean no disparaging reflection on such persons; they are as necessary to make up a world as cabbages to make up a garden; the great, healthy principles of cheerfulness and animal life seem to exist in them in the gross; they are wedges and ingots of solid, contented vitality.

A boy's growth.

"Oh, you go 'long, Massa Marvin; ye 'll live to count dat ar' boy for de staff o' yer old age yit, now I tell ye; got de makin' o' ten or'nary men in him; kittles dat's full allers will bile over; good yeast will blow at de cork, — lucky ef it don't bust de bottle. Tell ye, der's angels hes der hooks in sich, an' when de Lord wants him, dey 'll haul him in safe an' sound."

Will-power.

"Law me! what's de use? I 'se set out to b'liebe de Catechize, an' I 'se gwine to b'liebe it, so!"

The world's injustice.

"But, Marie, how unjust is the world! how unjust both in praise and blame."

OLDTOWN FOLKS.

Selfish love. These dear, good souls who wear their life out for you, have they not a right to scold you, and dictate to you, and tie up your liberty, and make your life a burden to you? If they have not, who has? If you complain, you break their worthy old hearts. They insist on the privilege of seeking your happiness by thwarting you in everything you want to do, and putting their will instead of yours in every step of your life.

Expressive silence. Aunt Lois, as I have often said before, was a good Christian, and held it her duty to govern her tongue. True, she said many sharp and bitter things; but nobody but herself and her God knew how many more she would have said had she not reined herself up in conscientious silence. But never was there a woman whose silence could express more contempt and displeasure than hers. You could feel it in the air about you, though she never said a word. You could feel it in the rustle of her dress, in the tap of her heels over the floor, in the occasional flash of her sharp black eye. She was like a thunder-cloud, whose quiet is portentous, and from which you every moment expect a flash or an explosion.

Power of a tone. That kind of tone which sounds so much like a blow that one dodges one's head involuntarily.

Making the best of it. "There's no use in such talk, Lois: what's done's done; and if the Lord let it be done, we may. We can't always make people do as we would. There's no use in being dragged through the world like a dog under a cart, hanging back and yelping. What we must do, we may as well do willingly, — as well walk as be dragged."

Influence of heredity and association. It is strange that no human being grows up who does not so intertwist in his growth the whole idea and spirit of his day, that rightly to dissect out his history would require one to cut to pieces and analyze society, law, religion, the metaphysics, and the morals of his time; and, as all things run back to those of past days, the problem is still further complicated. The humblest human being is the sum total of a column of figures which go back through centuries before he was born.

Personal magnetism. Supposing a man is made like an organ, with two or three banks of keys, and ever so many stops, so that he can play all sorts of tunes on himself; is it being a hypocrite with each person to play precisely the tune, and draw out exactly the stop, which he knows will make himself agreeable and further his purpose?

Physical good humor.
That charming gift of physical good humor, which is often praised as a virtue in children and in grown people, but which is a mere condition of the animal nature.

SAM LAWSON'S STORIES.

Effect of sinning.
"Ye know sinnin' will always make a man leave prayin'."

Scepticism.
"You look at the folks that's allers tellin' you what they don't believe, — they don't believe this, an' they don't believe that, — an' what sort o' folks is they? Why, like yer Aunt Lois, sort o' stringy an' dry. There ain't no 'sorption got out o' not believin' nothin'."

Life.
"That 'are 's jest the way folks go all their lives, boys. It's all fuss, fuss, and stew, stew, till ye get somewhere; an' then it's fuss, fuss, an' stew, stew, to get back again; jump here an' scratch your eyes out, an' jump there an' scratch 'em in again, — that 'are 's life."

PEARL OF ORR'S ISLAND.

Life as a play.
There are those people who possess a peculiar faculty of mingling in the affairs of this life as spectators as well as actors.

It does not, of course, suppose any coldness of nature or want of human interest or sympathy, — nay, it often exists more completely with people of the tenderest human feeling. It rather seems to be a kind of distinct faculty working harmoniously with all the others; but he who possesses it needs never to be at a loss for interest or amusement; he is always a spectator at a tragedy or a comedy, and sees in real life a humor and a pathos beyond anything he can find shadowed in books.

A child-like nature. Mrs. Pennel had one of those natures, gentle, trustful, and hopeful, because not very deep; she was one of the little children of the world, whose faith rests on childlike ignorance, and who know not the deeper needs of deeper natures; such see only the sunshine, and forget the storm.

Unintended hurts. All that there was developed of him, at present, was a fund of energy, self-esteem, hope, courage, and daring, the love of action, life, and adventure; his life was in the outward and present, not in the inward and reflective; he was a true ten-year-old boy, in its healthiest and most animal perfection. What she was, the small pearl with the golden hair, with her frail and high-strung organization, her sensitive nerves, her half-spiritual fibres, her ponderings, and marvels, and dreams, her power of love and yearning for self-devotion, our reader may, perhaps, have seen. But if ever two children, or two grown people, thus organized, are thrown into intimate relations, it follows, from the very laws of their being, that one must hurt the other, simply by being itself; one must always hunger for what the other has not to give.

Real love. "I always thought that my wife must be one of the sort of women who pray."

"And why?" said Mara, in surprise.

"Because I need to be loved a great deal, and it is only that kind who pray who know how to love *really*."

LITTLE FOXES.

Difficulty of self-knowledge. It is astonishing how much we think about ourselves, yet to how little purpose; how very clever people will talk and wonder about themselves and each other, not knowing how to use either themselves or each other, — not having as much practical philosophy in the matter of their own character and that of their friends as they have in respect to the screws of their gas-fixtures or the management of their water-pipes.

Reserve not understood. There are in every family circle individuals whom a certain sensitiveness of nature inclines to quietness and reserve; and there are very well-meaning families where no such quietness and reserve is possible. Nobody can be let alone, nobody may have a secret, nobody can move in any direction, without a host of inquiries and comments: " Who is your letter from? Let 's see." — " My letter is from So-and-so." — " He writing to you! I did n't know that. What 's he writing about? " — "Where did you go yesterday? What did you buy? What did you give for it? What are you going to do with it?" — " Seems to me that 's an odd way to do. I should n't do so." — " Look here, Mary; Sarah 's going to have a dress of silk tissue this spring. Now I think they 're too dear, don't you? "

I recollect seeing in some author a description of a true gentleman, in which, among other things, he was characterized as the man that asks the fewest questions. This trait of refined society might be adopted into home-life in a far greater degree than it is, and make it far more agreeable.

If there is perfect unreserve and mutual confidence, let it show itself in free communications coming unsolicited. It may fairly be presumed that, if there is anything our intimate friends wish us to know, they will tell us of it, and that when we are in close and confidential terms with persons, and there are topics on which they do not speak to us, it is because for some reason they prefer to keep silence concerning them; and the delicacy that respects a friend's silence is one of the charms of life.

Shyness of love. It comes far easier to scold our friend in an angry moment than to say how much we love, honor, and esteem him in a kindly mood. Wrath and bitterness speak themselves and go with their own force; love is shame-faced, looks shyly out of the window, lingers long at the door-latch.

Throwing away happiness. For the contentions that loosen the very foundations of love, that crumble away all its fine traceries and carved work, about what miserable, worthless things do they commonly begin! A dinner underdone, too much oil consumed, a newspaper torn, a waste of coal or soap, a dish broken! — and for this miserable sort of trash, very good, very generous, very religious people will sometimes waste and throw away by double-handfuls the very thing for which houses are built and all the paraphernalia of a home established, — *their happiness*. Better cold coffee, smoky tea, burnt meat, better any inconvenience, any loss, than a loss of *love;* and nothing so surely turns away love as constant fault-finding.

Morbid feelings. There is *fretfulness*, a mizzling, drizzling rain of discomforting remark; there is *grumbling*, a northeast snowstorm that never clears; there is *scolding*, the thunder-storm with lightning and hail. All these are worse than useless; they are positive *sins*, by whomsoever indulged, — sins as great and real as many that are shuddered at in polite society. All these are for the most part but the venting on our fellow-beings of morbid feelings resulting from dyspepsia, over-taxed nerves, or general ill-health.

HOUSE AND HOME PAPERS.

Love of a bargain. Milton says that the love of fame is the last infirmity of noble minds. I think he had not rightly considered the subject. I believe that last infirmity is the love of getting things cheap! Understand me, now. I don't mean the love of getting cheap things, by which one understands showy, trashy, ill-made, spurious articles, bearing certain apparent resemblances to better things. All really sensible people are quite superior to that sort of cheapness. But those fortunate accidents which put within the power of a man things really good and valuable for half or a third of their value, what mortal virtue and resolution can withstand ?

Warning for mothers. Mothers who throw away the key of their children's hearts in childhood sometimes have a sad retribution. As the children never were considered when they were little and helpless, so they do not consider when they are strong and powerful.

Careful observation. I think the best things on all subjects in this world of ours are said, not by the practical workers, but by the careful observers.

THE CHIMNEY CORNER.

Looking through blue glasses. Friend Theophilus was born on the shady side of Nature, and endowed by his patron saint with every grace and gift which can make a human creature worthy and available, except the gift of seeing the bright side of things. His bead-roll of Christian virtues includes all the graces of the spirit except hope; and so, if one wants to know exactly the flaw, the defect, the doubtful side, and to take into account all the untoward possibilities of any person, place, or thing, he had best apply to friend Theophilus. He can tell you just where and how the best-laid scheme is likely to fail, just the screw that will fall loose in the smoothest working machinery, just the flaw in the most perfect character, just the defect in the best written book, just the variety of thorn that must accompany each particular species of rose.

Châteaux en Espagne. Rudolph is another of the *habitués* of our chimney corner, representing the order of young knighthood in America, and his dreams and fancies, if impracticable, are always of a kind to make every one think him a good fellow. He who has no romantic dreams at twenty-one will be a horribly dry peascod at fifty; therefore it is that I gaze reverently at all Rudolph's châteaux in Spain, which want nothing to complete them except solid earth to stand on.

Care inevitable to human nature. The fact is that care and labor are as much correlated to human existence as shadow is to light; there is no such thing as excluding them from any mortal lot. You may make a canary-bird or a gold-fish live in absolute contentment without a care or labor, but a human being you cannot. Human beings are restless and active in their very nature, and will do something, and that something will prove a care, a labor, and a fatigue, arrange it how you will. As long as there is anything to be desired and not yet attained, so long its attainment will be attempted; so long as that attainment is doubtful or difficult, so long will there be care and anxiety.

THE MAYFLOWER.

"Cuteness." He possessed a great share of that characteristic national trait so happily denominated "cuteness," which signifies an ability to do everything without trying, to know everything without learning, and to make more use of one's *ignorance* than other people do of their knowledge.

Making people like us. It sometimes goes a great way towards making people like us to take it for granted that they do already.

A common mode of reasoning. She therefore repeated over exactly what she said before, only in a much

louder tone of voice, and with much more vehe-
ment forms of asseveration, — a mode of reasoning which, if not entirely logical, has at least the
sanction of very respectable authorities among
the enlightened and learned.

Danger in apparent safety. There is no point in the history of re-
form, either in communities or individ-
uals, so dangerous as that where danger
seems entirely past. As long as a man thinks
his health failing, he watches, he diets, and will
undergo the most heroic self-denial ; but let him
once set himself down as cured, and how readily
does he fall back to one soft, indulgent habit
after another, all tending to ruin everything that
he has before done !

Self-decep-tion. How strange that a man may appear
doomed, given up, and lost, to the eye
of every looker-on, before he begins to suspect
himself !

Convenient duties. What would people do if the convenient
shelter of duty did not afford them a
retreat in cases where they are disposed to change
their minds ?

Too much heart. A man can sometimes become an old
bachelor because he has *too much* heart,
as well as too little.

Privileged truth-tellers. These privileged truth-tellers are quite a
necessary of life to young ladies in the

full tide of society, and we really think it would be worth while for every dozen of them to unite to keep a person of this kind on a salary for the benefit of the whole.

Two kinds of frankness. There is one kind of frankness which is the result of perfect unsuspiciousness, and which requires a measure of ignorance of the world and of life; this kind appeals to our generosity and tenderness. There is another which is the frankness of a strong but pure mind, acquainted with life, clear in its discrimination and upright in its intention, yet above disguise or concealment; this kind excites respect. The first seems to proceed simply from impulse, the second from impulse and reflection united; the first proceeds, in a measure, from ignorance, the second from knowledge; the first is born from an undoubting confidence in others, the second from a virtuous and well-grounded reliance on one's self.

PINK AND WHITE TYRANNY.

Genial and ungenial natures. There are people who, wherever they move, freeze the hearts of those they touch, and chill all demonstration of feeling; and there are warm natures, that unlock every fountain, and bid every feeling gush forth.

Power of beauty. "Oh, nonsense! now, John, don't talk humbug. I 'd like to see *you* following goodness when beauty is gone. I 've known lots of plain old maids that were perfect saints and angels ; yet men crowded and jostled by them to get at the pretty sinners. I dare say now," she added, with a bewitching look over her shoulder at him, " you 'd rather have me than Miss Almira Carraway, — had n't you, now ? "

Growing alike. "The thing with you men," said Grace, " is that you want your wives to see with your eyes, all in a minute, what has got to come with years and intimacy, and the gradual growing closer and closer together. The husband and wife, of themselves, drop many friendships and associations that at first were mutually distasteful, simply because their tastes have grown insensibly to be the same."

DEACON PITKIN'S FARM.

A New England woman. Diana Pitkin was like some of the fruits of her native hills, full of juices which tend to sweetness in maturity, but which, when not quite ripe, have a pretty decided dash of sharpness. There are grapes that require a frost to ripen them, and Diana was somewhat akin to these.

AGNES OF SORRENTO.

Acceptable advice. Then he had given her advice which exactly accorded with her own views; and such advice is always regarded as an eminent proof of sagacity in the giver.

Dual nature. But, reviewing his interior world, and taking a survey of the work before him, he felt that sense of a divided personality which often becomes so vivid in the history of individuals of strong will and passion. It seemed to him that there were two men within him : the one turbulent, passionate, demented; the other vainly endeavoring, by authority, reason, and conscience, to bring the rebel to subjection. The discipline of conventual life, the extraordinary austerities to which he had condemned himself, the monotonous solitude of his existence, all tended to exalt the vivacity of the nervous system, which in the Italian constitution is at all times disproportionately developed ; and when those weird harp-strings of the nerves are once thoroughly unstrung, the fury and tempest of the discord sometimes utterly bewilders the most practiced self-government.

Power of an honest character. " Son, it is ever so," replied the monk. " If there be a man that cares neither for duke nor emperor, but for God alone, then dukes and emperors will give more for his good word than for a whole dozen of common priests."

Relation of age to youth. "We old folks are twisted and crabbed and full of knots with disappointment and trouble, like the mulberry-trees that they keep for vines to run on."

UNCLE TOM'S CABIN.

Persistence. "Dis yer matter 'bout persistence, feller-niggers," said Sam, with the air of one entering into an abstruse subject, "dis yer 'sistency 's a thing what ain't seed into very clar by most anybody. Now, yer see, when a feller stands up for a thing one day and night, de contrar' de next, folks ses (an' naturally enough dey ses), Why, he ain't persistent — hand me dat ar' bit o' corn-cake, Andy. But let 's look inter it. I hope the gent'lmen and de fair sex will scuse my usin' an or'nary sort o' 'parison. Here! I 'm a tryin' to get top o' der hay. Wal, I puts up my larder dis yer side; 't ain't no go; — den, 'cause I don't try dere no more, but puts my larder right de contrar' side, ain't I persistent? I 'm persistent in wantin' to get up which ary side my larder is; don't ye see, all on ye?"

The negro love of beauty. The negro, it must be remembered, is an exotic of the most gorgeous and superb countries of the world, and he has, deep in his heart, a passion for all that is splendid, rich, and fanciful; a passion which, rudely indulged by an untrained taste, draws on him the ridicule of the colder and more correct white race.

Effect of harshness. The ear that has never heard anything but abuse is strangely incredulous of anything so heavenly as kindness.

"Blessings brighten as they take their flight." Marie was one of those unfortunately constituted mortals, in whose eyes whatever is lost and gone assumes a value which it never had in possession. Whatever she had she seemed to survey only to pick flaws in it; but once fairly away, there was no end to her valuation of it.

DRED.

Speaking as a friend. "Now, Miss Nina, I want to speak as a friend."

"No, you sha'n't; it is just what people say when they are going to say something disagreeable. I told Clayton, once for all, that I would n't have him speak as a friend to me."

'Scuses. "Ah, lots of 'scuses I keeps! I tell you now, 'scuses is excellent things. Why, 'scuses is like dis yer grease dat keeps de wheels from screaking. Lord bless you, de whole world turns 'round on 'scuses. Whar de world be if everybody was such fools to tell de raal reason for everyting they are gwine for to do, or ain't gwine for to!"

Use of a chatterbox. Every kind of creature has its uses, and there are times when a lively, unthinking chatterbox is a perfect godsend. Those unperceiving people who never notice the embarrassment of others, and who walk with the greatest facility into the gaps of conversation, simply because they have no perception of any difficulty there, have their hour; and Nina felt positively grateful to Mr. Carson for the continuous and cheerful rattle which had so annoyed her the day before.

Good and evil inseparable. It is our fatality that everything that does good must do harm. It is the condition of our poor, imperfect life here.

"Streaked men." "But den, you see, honey, der's some folks der's *two* men in 'em, — one is a good one, and t'oder is very bad. Wal, dis yer was jest dat sort. . . . He was one of dese yer streaked men, dat has dreful ugly streaks; and, some of dem times, de Lord only knows what he won't do."

First steps. There is something in the first essay of a young man, in any profession, like the first launching of a ship, which has a never-ceasing hold on human sympathies.

From different standpoints. There is no study in human nature more interesting than the aspects of the same subject seen in the points of view of different characters.   One might almost imagine that there were no such thing as absolute truth, since a change of situation or temperament is capable of changing the whole force of an argument.

Fine natures perverted. As good wine makes the strongest vinegar, so fine nature perverted makes the worst vice.

SUNNY MEMORIES OF FOREIGN LANDS.

Lost confidence. There are some people who involve in themselves so many of the elements which go to make up our confidence in human nature generally, that to lose confidence in them seems to undermine our faith in human virtue.

Wit. Truly, wit, like charity, covers a multitude of sins.   A man who has the faculty of raising a laugh in this sad, earnest world is remembered with indulgence and complacency.

Value of ready expression. But so it always is.   The man who has exquisite gifts of expression passes for more, popularly, than the man with great and grand ideas, who utters but imperfectly.

POGANUC PEOPLE.

Opinion-ated peo-ple. Miss Debby was one of those human beings who carry with them the apology for their own existence. It took but a glance to see that she was one of those forces of nature which move always in straight lines, and which must be turned out for if one wishes to avoid a collision. All Miss Debby's opinions had been made up, catalogued, and arranged at a very early period of life, and she had no thought of change. She moved in a region of certainties, and always took her own opinions for granted with a calm supremacy altogether above reason. Yet there was all the while about her a twinkle of humorous consciousness, a vein of original drollery, which gave piquancy to the brusqueness of her manner, and prevented people from taking offence.

Difficulty of confes-sion. It is curious that men are not generally ashamed of any form of anger, wrath, or malice ; but of the first step towards a nobler nature, — the confession of a wrong, — they are ashamed.

LITTLE PUSSY WILLOW.

Animal spirits. When people work hard all day, and have a good digestion, it is not necessary that a thing should be very funny to make them laugh tremendously.

First false step. Boys, and men too, sometimes, by a single step, and that step taken in a sudden hurry of inconsideration, get into a network of false positions, in which they are very uneasy and unhappy, but live along from day to day seeing no way out.

QUEER LITTLE PEOPLE.

Marks of genius. "Depend upon it, my dear," said Mrs. Nut-cracker, solemnly, "that fellow must be a genius."

"Fiddlestick on his genius!" said old Mr. Nut-cracker; "what does he do?"

"Oh, nothing, of course; that's one of the first marks of genius. Geniuses, you know, never can come down to common life."

A busy-body. Old Mother Magpie was about the busiest character in the forest. But you must know that there is a great difference between being busy and being industrious. One may be very busy all the time, and yet not in the least industrious; and this was the case with Mother Magpie.

She was always full of everybody's business but her own, — up and down, here and there, everywhere but in her own nest, knowing every one's affairs, telling what everybody had been doing or ought to do, and ready to cast her advice *gratis* at every bird and beast of the woods.

A DOG'S MISSION.

Broken idols. Do you, my brother, or grown-up sister, ever do anything like this? Do your friendships and loves ever go the course of our Charley's toy? First, enthusiasm; second, satiety; third, discontent; then picking to pieces; then dropping and losing! How many idols are in your box of by-gone playthings? And may it not be as well to suggest to you, when you find flaws in your next one, to inquire before you pick to pieces whether you can put together again, or whether what you call defect is not a part of its nature? A tin locomotive won't draw a string of parlor chairs, by any possible alteration, but it may be very pretty for all that it was made for. Charley and you might both learn something from this.

MY WIFE AND I.

Soul-language. "There are people in this world who don't understand each other's vernacular. Papa and I could no more discuss any question of the inner life together than if he spoke Chickasaw and I spoke French."

Characters worth exploring. It is a charming thing, in one's rambles, to come across a tree, or a flower, or a fine bit of landscape that we can think of afterwards, and feel richer for its being in the world. But it is more, when one is in a strange place, to come across a man that you feel thoroughly persuaded is, somehow or other, morally and intellectually worth exploring. Our lives tend to become so hopelessly commonplace, and the human beings we meet are generally so much one just like another, that the possibility of a new and peculiar style of character in an acquaintance is a most enlivening one.

Unsuspected danger. The man who has begun to live and work by artificial stimulant never knows where he stands, and can never count upon himself with any certainty. He lets into his castle a servant who becomes the most tyrannical of masters. He may resolve to turn him out, but will find himself reduced to the condition in which he can neither do with nor without him.

In short, the use of stimulant to the brain power brings on a disease in whose paroxysms a man is no more his own master than in the ravings of fever, a disease that few have the knowledge to understand, and for whose manifestations the world has no pity.

Heredity. Out of every ten young men who begin the use of stimulants as a social exhilaration, there are perhaps five in whose breast lies coiled

up and sleeping this serpent, destined in after
years to be the deadly tyrant of their life — this
curse, unappeasable by tears, or prayers, or
agonies — with whom the struggle is like that of
Laocoön with the hideous python, yet songs and
garlands and poetry encircle the wine-cup, and
ridicule and contumely are reserved for him who
fears to touch it.

Personality. We are all familiar with the fact that
there are some people who, let them sit still as
they may, and conduct themselves never so
quietly, nevertheless impress their personality on
those around them, and make their presence felt.

WE AND OUR NEIGHBORS.

Friendly gossip. A great deal of good sermonizing, by
the way, is expended on gossip, which
is denounced as one of the seven deadly sins of
society ; but, after all, gossip has its better side ;
if not a Christian grace, it certainly is one of
those weeds which show a good warm soil.

The kindly heart, that really cares for everything human it meets, inclines toward gossip in a
good way. Just as a morning-glory throws out
tendrils, and climbs up and peeps cheerily into
your window, so a kindly gossip can't help watching the opening and shutting of your blinds, and
the curling smoke from your chimney.

Persist-
ency. If you *will* have your own way, and
persist in it, people *have to* make up
with you.

Right side
of human
nature. Human nature is always interesting, if
one takes it right side out.

Human
nettles. It is rather amusing to a general looker-
on in this odd world of ours to contrast
the serene, cheerful good faith with which these
constitutionally active individuals go about criti-
cising, and suggesting, and directing right and
left, with the dismay and confusion of mind they
leave behind them wherever they operate.

They are often what the world calls well-
meaning people, animated by a most benevolent
spirit, and have no more intention of giving
offence than a nettle has of stinging. A large,
vigorous, well-growing nettle has no consciousness
of the stings it leaves in the delicate hands that
have been in contact with it; it has simply acted
out its innocent and respectable nature as a
nettle. But a nettle armed with the power of
locomotion on an ambulatory tour, is something
the results of which may be fearful to contem-
plate.

Flaws in
gems. Ideal heroes are not plentiful, and there
are few gems that don't need rich setting.

Impossi-
bility of
evading
trouble.
People who hate trouble generally get a good deal of it. It's all very well for a gentle, acquiescent spirit to be carried through life by *one* bearer. But when half a dozen bearers quarrel and insist on carrying one opposite ways, the more facile the spirit, the greater the trouble.

Righteous-
ness
through re-
pentance.
Perhaps there is never a time when man or woman has a better chance, with suitable help, of building a good character, than just after a humiliating fall which has taught the sinner his own weakness, and given him a sad experience of the bitterness of sin.

Nobody wants to be sold under sin, and go the whole length in iniquity ; and when one has gone just far enough in wrong living to perceive in advance all its pains and penalties, there is often an agonized effort to get back to respectability, like the clutching of the drowning man for the shore. The waters of death are cold and bitter, and nobody wants to be drowned.

"I told
you so."
" Whence is the feeling of satisfaction which we have when things that we always said we knew turn out just as we predicted ? Had we really rather our neighbor would be proved a thief and a liar than to be proved in a mistake ourselves ? Would we be willing to have somebody topple headlong into destruction for the sake of being able to say, 'I told you so ' ? "

Gossip.   In fact, the gossip plant is like the grain of mustard-seed, which, though it be the least of all seeds, becometh a great tree, and the fowls of the air lodge in its branches, and chatter mightily there at all seasons.

CHAPTER III.

WOMAN.

THE MINISTER'S WOOING.

Woman as a Gospel. "You girls and women don't know your power. Why, Mary, you are a living Gospel. You have always had a strange power over us boys. You never talked religion much, but I have seen high fellows come away from being with you as still and quiet as one feels when he goes into a church. I can't understand all the hang of predestination and moral ability, and natural ability, and God's efficiency, and man's agency, which Dr. Hopkins is so engaged about; but I can understand *you*, — *you* can do me good."

Holiness of woman. "But do you remember you told me once that, when the snow first fell, and lay so dazzling and pure and soft all about, you always felt as if the spreads and window curtains, that seemed white before, were not clean? Well, it's just like that with me. Your presence makes me feel that I am not pure, — that I am low and unworthy, — not worthy to touch the hem of your garment. Your good Dr. Hopkins

spent a whole half day, the other Sunday, trying
to tell us about the beauty of holiness; and he
cut, and pared, and peeled, and sliced, and told
us what it was n't; and what was *like it*, and
was n't; and then he built up an exact definition,
and fortified and bricked it up all round; and I
thought to myself that he 'd better tell 'em to
look at Mary Scudder, and they 'd understand all
about it."

Woman en- Do you remember, at Niagara, a little
nobled by
man's love. cataract on the American side, which
throws its silver, sheeny veil over a cave
called the Grot of Rainbows? Whoever stands
on a rock in that grotto sees himself in the cen-
tre of a rainbow-circle, above, below, around.
In like manner, merry, chatty, positive, busy,
house-wifely Katy saw herself standing in a rain-
bow-shrine in her lover's inner soul, and liked to
see herself so. A woman, by-the-bye, must be
very insensible, who is not moved to come upon
a higher plane of being herself, by seeing how
undoubtingly she is insphered in the heart of a
good and noble man. A good man's faith in
you, fair lady, if you ever have it, will make
you better and nobler even before you know it.

Power of It is only now and then that a matter-
real love. of-fact woman is sublimated by a real
love; but if she is, it is affecting to see how impossible it is for death to quench it.

Woman's
veneration.

If women have one weakness more marked than another, it is towards veneration. They are born worshippers. . . . The fact is, women are burdened with fealty, faith, and reverence, more than they know what to do with; they stand like a hedge of sweet-peas, throwing out fluttering tendrils everywhere for something high and strong to climb by, — and when they find it, be it ever so rough in the bark, they catch upon it. And instances are not wanting of those who have turned away from the flattery of admirers to prostrate themselves at the feet of a genuine hero who never wooed them, except by heroic deeds and the rhetoric of a noble life.

Mother-
love for
a son.

None of the peculiar developments of the female nature have a more exquisite vitality than the sentiment of a frail, delicate, repressed, timid woman, for a strong, manly, generous son. There is her ideal expressed; there is the outspeaking and outacting of all she trembles to think, yet burns to say or do; here is the hero that shall speak for her, the heart into which she has poured hers, and that shall give to her tremulous and hidden aspirations a strong and victorious expression. " I have gotten a *man* from the Lord," she says to herself, and each outburst of his manliness, his vigor, his self-confidence, his superb vitality, fills her with a strange, wondering pleasure, and she has a secret tenderness and pride even in his wilful-

ness and waywardness. . . . First love of womanhood is something wonderful and mysterious,
— but in this second love it rises again, idealized
and refined; she loves the father and herself
united and made one in this young heir of life
and hope.

Mothers' inconsid-erateness. But even mothers who have married
for love themselves somehow so blend
a daughter's existence with their own as
to conceive that she must marry their love and
not her own.

Repression. Her large brown eyes had an eager joy
in them when Mary entered; but they seemed
to calm down again, and she received her only
with that placid, sincere air which was her habit.
Everything about this woman showed an ardent
soul, repressed by timidity and by a certain
dumbness in the faculties of outward expression;
but her eyes had, at times, that earnest, appeal-
ing language which is so pathetic in the silence
of inferior animals. One sometimes sees such
eyes, and wonders whether the story they intimate will ever be spoken in mortal language.

Woman's instinctive silence. Ah, that silence! Do not listen to hear
whom a woman praises, to know where
her heart is! do not ask for whom she
expresses the most earnest enthusiasm! but if
there be one she once knew well whose name she
never speaks, — if she seems to have an instinct

to avoid every occasion of its mention, — if
when you speak, she drops into silence and
changes the subject, — why, look there for something! just so, when going through deep mead-
ow-grass, a bird flies ostentatiously up before you,
you may know her nest is not there, but far off,
under distant tufts of fern and buttercup, through
which she has crept with a silent flutter in her
spotted breast, to act her pretty little falsehood
before you.

Idle talk.   When Mrs. Twitchel began to talk, it
flowed a steady stream, as when one turns a
faucet, that never ceases running till some hand
turns it back again.

Reverence the basis of faith.   "Who cares?" said Candace, — "gen-
erate or unregenerate, it's all one to
me; I believe a man dat *acts* as he does.
Him as stands up for de poor, — him as pleads
for de weak, — he's my man. I'll believe straight
through anyting he's a mind to put at me."

Mothers' intuition.   Most mothers are instinctive philosophers. No treatise on the laws of
nervous fluids could have taught Mrs. Scudder a
better *rôle* for this morning, than her tender
gravity, and her constant expedients to break
and ripple, by changing employments, that deep,
deadly undercurrent of thoughts which she feared
might undermine her child's life.

OLDTOWN FOLKS.

Woman's nature. It is a man's nature to act, to do, and when nothing can be done, to forget. It is a woman's nature to hold on to what can only torture, and live all her despairs over. Women's tears are their meat; men find the diet too salt, and won't take it.

Using knowledge. "My forte lies in picking knowledge out of other folks and using it," said Tina, joyously. "Out of the least little bit of ore that you dig up, I can make no end of gold-leaf."

Mothers' work. "Ain't the world hard enough without fightin' babies, I want to know? I hate to see a woman that don't want to rock her own baby, and is contriving ways all the time to shirk the care of it. Why, if all the world was that way, there would be no sense in Scriptur'. 'As one whom his mother comforteth, so will I comfort you,' the Bible says, takin' for granted that mothers were made to comfort children and give them good times when they are little."

The mother in every woman. "There's no saying," said Miss Mehitable, "you never know what you may find in the odd corners of an old maid's heart, when you fairly look into them. There are often unused hoards of maternal affec-

tion enough to set up an orphan asylum; but it 's
like iron filings and a magnet, — you must try
them with a live child, and if there is anything
in 'em, you 'll find it out. That little object," she
said, looking over her shoulder at Tina, "made
an instant commotion in the dust and rubbish of
my forlorn old garret, brought to light a deal
that I thought had gone to the moles and the
bats long ago. She will do me good, I can feel,
with her little pertnesses, and her airs and fancies. If you could know how chilly and lone-
some an old house gets sometimes, particularly in
autumn, when the equinoctial storm is brewing !
A lively child is a godsend, even if she turns the
whole house topsy-turvy."

Individ-
uality.
Tina had one of those rebellious heads
of curls that every breeze takes liberties
with, and that have to be looked after, and
watched, and restrained. Esther's satin bands
of hair could pass through a whirlwind and not
lose their gloss. It is curious how character
runs even to the minutest thing, — the very hairs
of our head are numbered by it, — Esther,
always and in everything self-poised, thoughtful,
reflective; Tina, the child of every wandering
influence, tremulously alive to every new excitement, a wind-harp for every air of heaven to
breathe upon.

A woman's
view.
"The fact is, a man never sees a subject thoroughly till he sees what a

woman will think of it, for there is a woman's view of every subject, which has a different shade from a man's view, and that is what you and I have insensibly been absorbing in all our course hitherto."

PEARL OF ORR'S ISLAND.

Neighbor's influence. Duty is never more formidable than when she gets on the cap and gown of a neighbor.

Reserve. But it was not the little maiden's way to speak when anything thwarted or hurt her, but rather to fold all her feelings and thoughts inward, as some insects, with fine gauzy wings, draw them under a coat of horny concealment.

True courage. That kind of innocent hypocrisy which is needed as a staple in the lives of women who bridge a thousand awful chasms with smiling, unconscious looks, and walk, singing and scattering flowers, over abysses of fear, when their hearts are dying within them.

The reserve power of quietness. Pliable as she was to all outward appearances, the child had her own still, interior world, where all her little notions and opinions stood up, crisp and fresh, like flowers that grow in cool, shady places. If anybody too rudely assailed a thought or suggestion she put forth, she drew it back again into this quiet, inner chamber, and went on. Reader, there are some women of this habit; there is no independence and pertinacity of opinion like that of those seemingly soft, quiet creatures, whom it is so easy to silence, and so difficult to convince. Mara, little and unformed as she yet was, belonged to the race of those spirits to whom is deputed the office of the angel in the Apocalypse, to whom was given the golden rod which measured the new Jerusalem. Infant though she was, she had ever in her hands that invisible measuring rod, which she was laying to the foundations of all actions and thoughts. There may, perhaps, come a time when the saucy boy, who now steps so superbly, and predominates so proudly in virtue of his physical strength and daring, will learn to tremble at the golden measuring rod held in the hand of a woman.

Sweetness. " She 's got sweet ways and kind words for everybody, and it 's as good as a psalm to look at her."

Woman's life within. No man — especially one that is living a rough, busy, out-of-doors life — can form the slightest conception of that veiled and secluded life which exists in the heart of a sensitive woman, whose sphere is narrow, whose external diversions are few, and whose mind, therefore, acts by a continual introversion upon itself. They know nothing how their careless

words and actions are pondered and turned again
in weary, quiet hours of fruitless questioning.
What did he mean by this? and what did he
mean by that? — while he, the careless buffalo,
meant nothing, or has forgotten what it was, if
he did.

Girls' con-
fidences. "Come, now, can't you jest tramp
over to Pennel's and tell Sallie I want
her?"

"Not I, mother. There ain't but two gals in
two miles square here, an' I ain't a-goin' to be
the feller to shoo 'em apart. What 's the use o'
bein' gals, an' young, an' pretty, if they can't get
together an' talk about their new gowns an' the
fellers? That ar 's what gals is for."

Maternal
element in
woman's
love. Her love for Moses had always had in
it a large admixture of that maternal
and care-taking element, which, in some
shape or other, qualifies the affection of woman
to man.

LITTLE FOXES.

Tact. Some women are endowed with a tact
for understanding human nature and guiding it.
They give a sense of largeness and freedom;
they find a place for every one, see at once what
every one is good for, and are inspired by nature
with the happy wisdom of not wishing or asking
of any human being more than that human being

was made to give. They have the portion in due season for all: a bone for the dog; catnip for the cat; cuttle-fish and hemp-seed for the bird; a book or review for their bashful literary visitor; lively gossip for thoughtless Miss Seventeen; knitting for grandmamma; fishing-rods, boats, and gunpowder for Young Restless, whose beard is just beginning to grow; — and they never fall into pets, because the canary-bird won't relish the dog's bone, or the dog eat canary-seed, or young Miss Seventeen read old Mr. Sixty's review, or young Master Restless take delight in knitting-work, or old grandmamma feel complacency in guns and gunpowder.

Again, there are others who lay the foundations of family life so narrow, straight, and strict, that there is room in them only for themselves and people exactly like themselves; and hence comes much misery.

Modern saints. Talk of hair-cloth shirts, and scourgings, and sleeping in ashes as a means of saintship! there is no need of them in our country. Let a woman once look at her domestic trials as her hair-cloth, her ashes, her scourges, — accept them, rejoice in them, smile and be quiet, silent, patient, and loving under them, — and the convent can teach her no more. She is a victorious saint.

HOUSE AND HOME PAPERS.

A help-
meet. My wife resembles one of those convex mirrors I have sometimes seen. Every idea I threw out, plain and simple, she reflected back upon me in a thousand little glitters and twinkles of her own; she made my crude conceptions come back to me in such perfectly dazzling performances that I hardly recognized them.

A true
home. How many, morally wearied, wandering, disabled, are healed and comforted by the warmth of a true home! When a mother has sent her son to the temptations of a distant city, what news is so glad to her heart as that he has found some quiet family where he visits often and is made to feel *at home?* How many young men have good women saved from temptation and shipwreck, by drawing them often to the sheltered corner by the fireside! The poor artist — the wandering genius who has lost his way in this world, and stumbles like a child among hard realities, — the many men and women, who, while they have houses, have no homes, — see from afar, in their distant, bleak life-journey, the light of a true home-fire, and if made welcome there, warm their stiffened limbs, and go forth stronger to their pilgrimage. Let those who have accomplished this beautiful and perfect work of divine art be liberal of its influence.

Let them not seek to bolt the doors and draw the curtains; for they know not, and will never know till the future life, of the good they may do by the ministration of this great charity of home.

THE CHIMNEY CORNER.

The art of home-mak-ing. She alone can keep the poetry and beauty of married life who has this poetry in her soul; who with energy and discretion can throw back and out of sight the sordid and disagreeable details which beset all human living, and can keep in the foreground that which is agreeable; who has enough knowledge of practical household matters to make unskilled and rude hands minister to her cultivated and refined tastes, and constitute her skilled brain the guide of unskilled hands. From such a home, with such a mistress, no sirens will seduce a man, even though the hair grow gray, and the merely physical charms of early days gradually pass away. The enchantment that was about her person alone in the days of courtship seems in the course of years to have interfused and penetrated the *home* which she has created, and which in every detail is only an expression of her personality. Her thoughts, her plans, her provident care, are everywhere; and the *home* attracts and holds by a thousand ties the heart which before marriage was held by the woman alone.

THE MAYFLOWER.

A perfect character. " Was she beautiful ? " you ask. I also will ask you one question : " If an angel from heaven should dwell in human form, and animate any human face, would not that face be lovely? It might not be *beautiful*, but would it not be lovely?" She was not beautiful except after this fashion.

How well I remember her, as she used sometimes to sit thinking, with her head resting on her hand, her face mild and placid, with a quiet October sunshine in her blue eyes, and an ever-present smile over her whole countenance. I remember the sudden sweetness of look when any one spoke to her; the prompt attention, the quick comprehension of things before you uttered them, the obliging readiness to leave for you whatever she was doing.

To those who mistake occasional pensiveness for melancholy, it might seem strange to say that my Aunt Mary was always happy. Yet she was so. Her spirits never rose to buoyancy, and never sunk to despondency. I know that it is an article in the sentimental confession of faith that such a character cannot be interesting. For this impression there is some ground. The placidity of a medium, commonplace mind is uninteresting, but the placidity of a strong and well governed one borders on the sublime. Mutability of emotion characterizes inferior orders of being;

but He who combines all interest, all excitement, all perfection, is "the same yesterday, to-day, and forever." And if there be anything sublime in the idea of an Almighty Mind, in perfect peace itself, and, therefore, at leisure to bestow all its energies on the wants of others, there is at least a reflection of the same sublimity in the character of that human being who has so quieted and governed the world within that nothing is left to absorb sympathy or distract attention from those around.

Such a woman was my Aunt Mary. Her placidity was not so much the result of temperament as of choice. She had every susceptibility of suffering incident to the noblest and most delicate constitution of mind ; but they had been so directed that, instead of concentrating thought on self, they had prepared her to understand and feel for others.

She was, beyond all things else, a sympathetic person, and her character, like the green in a landscape, was less remarkable for what it was in itself than for its perfect and beautiful harmony with all the coloring and shading around it.

Other women have had talents, others have been good; but no woman that ever I knew possessed goodness and talent in union with such an intuitive perception of feelings, and such a faculty of instantaneous adaptation to them. The most troublesome thing in this world is to be condemned to the society of a person who can never understand anything you say unless you say the whole of it, making your commas and periods as you go along; and the most desirable thing in the world is to live with a person who saves you all the trouble of talking by knowing just what you mean before you begin to speak.

Woman's moral influence. "That word *delicacy* is a charming cover-all in all these cases, Florence.

Now, here is a fine, noble-spirited young man, away from his mother and sisters, away from any family friend who might care for him, tempted, betrayed, almost to ruin, and a few words from you, said as a woman knows how to say them, might be his salvation. But you will coldly look on and see him go to destruction, because you have too much delicacy to make the effort — like the man that would not help his neighbor out of the water because he had never had the honor of an introduction."

" But, Edward, consider how peculiarly fastidious Elliot is — how jealous of any attempt to restrain and guide him."

" And just for that reason it is that *men* of his acquaintance cannot do anything with him. But what are you women made with so much tact and power of charming for, if it is not to do these very things that we cannot do? It is a delicate matter — true; and has not Heaven given to you a fine touch and a fine eye for just such delicate matters? Have you not seen, a thousand times, that what might be resented as an impertinent interference on the part of a man comes to us as a flattering expression of interest from the lips of a woman?"

PINK AND WHITE TYRANNY.

Selfishness. That kind of woman can't love. They are like cats, that want to be stroked and caressed, and to be petted, and to lie soft and warm; and they will purr to any one that will pet them, — that's all. As for love that leads to any self-sacrifice, they don't begin to know anything about it.

Intuition. Now Grace had that perfect intuitive knowledge of just what the matter was with her brother that women always have who have grown up in intimacy with a man. These fine female eyes see farther between the rough cracks and ridges of the oak-bark of manhood than men themselves.

DEACON PITKIN'S FARM.

The New England wife-mother. New England had of old times, and has still, perhaps, in her farm-houses, these women who seem from year to year to develop in the spiritual sphere as the bodily form shrinks and fades. While the cheek grows thin and the form spare, the will-power grows daily stronger; though the outer man perish, the inner

man is renewed day by day. The worn hand
that seems so weak yet holds every thread and
controls every movement of the most complex
family life, and wonders are daily accomplished
by the presence of a woman who seems little
more than a spirit. The New England wife-
mother was the one little jeweled pivot on which
all the wheel-work of the family moved.

Suppres-
sion.
It was not the first time that, wounded
by a loving hand in this dark struggle
of life, she had suppressed the pain of her own
hurt, that he that had wounded her might the
better forgive himself.

AGNES OF SORRENTO.

True
beauty.
"A beautiful face is a kind of psalm
which makes one want to be good."

Forcing a
daughter.
"After all, sister, what need of haste?
'T is a young bird yet. Why push it
out of the nest? When once it is gone you will
never get it back. Let the pretty one have her
little day to play and sing and be happy. Does
she not make this garden a sort of Paradise with
her little ways and her sweet words? Now, my
sister, these all belong to you; but, once she is
given to another, there is no saying what may
come. One thing only may you count on with
certainty: that these dear days when she is all

day by your side and sleeps in your bosom all
night are over, — she will belong to you no
more, but to a strange man who hath neither
toiled nor wrought for her, and all her pretty
ways and dutiful thoughts must be for him."

UNCLE TOM'S CABIN.

Beautiful old age. Her face was round and rosy, with a
healthful, downy softness, suggestive of
a ripe peach. Her hair, partially silvered by
age, was parted smoothly back from her high,
placid forehead, on which time had written no
inscription except " Peace on earth, good will to
men," and beneath shone a large pair of clear,
honest, loving, brown eyes ; you only needed to
look straight into them, to feel that you saw to
the bottom of a heart as good and true as ever
throbbed in woman's bosom. So much has been
said and sung of beautiful young girls, why don't
somebody wake up to the beauty of old women ?

Exaction. It is a great mistake to suppose that a
woman with no heart will be an easy creditor in
the exchange of affection. There is not on earth
a more merciless exactor of love from others
than a thoroughly selfish woman ; and the more
unlovely she grows, the more jealously and scru-
pulously she exacts love to the uttermost far-
thing.

PALMETTO LEAVES.

Character. A flower is commonly thought the emblem of a woman; and a woman is generally thought of as something sweet, clinging, tender, and perishable. But there are women flowers that correspond to the forest magnolia, — high and strong, with a great hold of root and a great spread of branches; and whose pulsations of heart and emotion come forth like these silver lilies that illuminate the green shadows of the magnolia forests.

POGANUC PEOPLE.

"Turn about." "Oh, land o' Goshen, Dolly, what do you mind them boys for?" said Nabby. "Boys is mostly hateful when girls is little; but we take our turns by and by," she said, with a complacent twinkle of her brown eyes. "I make them stand 'round, I bet ye, and you will when you get older."

MY WIFE AND I.

Woman's spiritual power. My mother was one of that class of women whose power on earth seems to be only the greater for being a spiritual and invisible one. The control of such women over men is like that of the soul over the body.

The body is visible, forceful, obtrusive, self-asserting; the soul, invisible, sensitive, yet with a subtile and vital power which constantly gains control, and holds every inch that it gains.

Orderliness. Like a true little woman, she seemed to have nerves through all her clothes, that kept them in order.

Woman and Christianity. The motherly instinct is in the hearts of all true women, and sooner or later the true wife becomes a mother to her husband; she guides him, cares for him, teaches him, and catechizes him, all in the nicest way possible. . . . As for the soul-life, I believe it is woman who holds faith in the world, — it is woman behind the wall, casting oil on the fire that burns brighter and brighter, while the devil pours on water; and you 'll never get Christianity out of the world while there 's a woman in it.

WE AND OUR NEIGHBORS.

Woman's mission. "That 's what you women are for — at least such women as you. It 's your mission to interpret differing natures — to bind, and blend, and unite."

Real conversation. That fine, skillful faculty of analysis and synthesis which forms the distinctive interest of feminine conversation.

CHAPTER IV.

CHILDREN.

THE MINISTER'S WOOING.

The odd one. One sometimes sees launched into a family circle a child of so different a nature from all the rest, that it might seem as if, like an aërolite, he had fallen out of another sphere.

OLDTOWN FOLKS.

Child's intensity. In childhood the passions move with a simplicity of action unknown to any other period of life, and a child's hatred and a child's revenge have an intensity of bitterness entirely unalloyed by moral considerations; and when a child is without an object of affection and feels itself unloved, its whole vigor of being goes into the channels of hate.

Child instinct. That instinctive sense by which children and dogs learn the discerning of spirits.

Childish antipathies.

Among the many unexplained and inexplicable woes of childhood are its bitter antagonisms, so perfectly powerless, but often so very decided, against certain of the grown people who control it. Perhaps some of us may remember respectable, well-meaning people, with whom in our mature years we live in perfect amity, but who in our childhood appear to us bitter enemies. Children are remarkably helpless in this respect, because they cannot choose their company and surroundings as grown people can; and are sometimes entirely in the power of those with whom their natures are so unsympathetic that they may almost be said to have a constitutional aversion to them.

Getting used to the world.

Nobody that has not suffered from such causes can tell the amount of torture that a child of a certain nervous formation undergoes in the mere process of getting accustomed to his body, to the physical forces of life, and to the ways and doings of that world of grown-up people who have taken possession of the earth before him, and are using it, and determined to go on using it, for their own behoof and convenience, in spite of his childish efforts to push in his little individuality, and seize his little portion of existence. He is at once laid hold upon by the older majority as an instrument to work out their views of what is fit and proper for himself and themselves; and if he proves a hard-working or creaking instrument, has the further capability of being rebuked and chastened for it.

Quiet children. I was one of those children who are all ear, — dreamy listeners, who brood over all that they hear, without daring to speak of it.

Individuality in children. He was one of those children who retreat into themselves and make a shield of quietness and silence in the presence of many people, while Tina, on the other hand, was electrically excited, waxed brilliant in color, and rattled and chattered with as fearless confidence as a cat-bird.

A child's philosophy. "But, Tina, mother always told us it was wicked to hate anybody. We must love our enemies."

"You don't love old Crab Smith, do you?"

"No, I don't; but I try not to hate him," said the boy. "I won't think anything about him."

"I can't help thinking," said Tina; "and when I think, I am so angry! I feel such a burning in here!" she said, striking her little breast; "it's just like fire."

"Then don't think about her at all," said the boy; "it is n't pleasant to feel that way. Think about the whip-poor-wills singing in the woods over there, — how plain they say it, don't they? — And the frogs all singing, with their little,

round, yellow eyes looking up out of the water;
and the moon looking down on us so pleasantly!
she seems just like mother!"

A child's
questions.
Is there ever a hard question in morals
that children do not drive straight at in
their wide-eyed questioning?

PEARL OF ORR'S ISLAND.

Holiness of
infancy.
The wise men of the east at the feet of
an infant, offering gifts, gold, frankincense, and myrrh, is just a parable of what goes
on in every house where there is a young child.
All the hard and the harsh, the common and the
disagreeable, is for the parents, — all the bright
and beautiful for their child.

Pure joy.　Childhood's joys are all pure gold.

Mischief.　"Of all the children that ever she see,
he beat all for finding out new mischief, — the
moment you make him understand he must n't
do one thing, he 's right at another."

Different
tempera-
ments.
"Mis' Pennel ought to be trainin' of
her up to work," said Mrs. Kittridge.
"Sally could oversew and hem when
she wa' n't more 'n three years old; nothin'
straightens out children like work. Mis' Pennel
she jest keeps that ar' child to look at."

"All children a'n't alike, Mis' Kittridge," said Miss Roxy, sententiously. "This 'un a'n't like your Sally. 'A hen and a bumble-bee can't be fetched up alike,' fix it how you will!"

Child's buoyancy. All the efforts of Nature, during the early years of a healthy childhood, are bent on effacing and obliterating painful impressions, wiping out from each day the sorrows of the last, as the daily tide effaces the furrows on the seashore.

Unseen dangers. Neither of them had known a doubt or a fear in that joyous trance of forbidden pleasure, which shadowed with so many fears the wiser and more far-seeing heads and hearts of the grown people; nor was there enough language yet in common between the two classes to make the little ones comprehend the risk they had run.

Perhaps our older brothers, in our Father's house, look anxiously out when we are sailing gayly over life's sea, over unknown depths, amid threatening monsters, but want words to tell us why what seems so bright is so dangerous.

Love of solitude. The island was wholly solitary, and there is something to children quite delightful in feeling that they have a little, lonely world all to themselves. Childhood is itself such an enchanted island, separated by mysterious depths from the main land of nature, life, and reality.

Fate.   But babies will live, all the more when everybody says it is a pity they should. Life goes on as inexorably in this world as death.

Sensitive   There are natures sent down into this
natures.   harsh world so timorous, sensitive, and helpless in themselves, that the utmost stretch of indulgence and kindness is needed for their development, — like plants which the warmest shelf of the green-house and the most watchful care of the gardener alone can bring into flower.

Child's   " It 's curious what notions chil'en will
reasoning. get in their heads," said Captain Kittridge. " They put this an' that together and think it over, an' come out with such queer things."

THE CHIMNEY CORNER.

A child's   The hearts of little children are easily
love.   gained, and their love is real and warm, and no true woman can become the object of it without feeling her own life made brighter.

THE MAYFLOWER.

A child's   But the feelings of grown-up children
longing for
sympathy. exist in the minds of little ones oftener than is supposed ; and I had, even at this early day, the same keen sense of all that touched

the heart wrong; the same longing for something which should touch it aright; the same discontent with latent, matter-of-course affection, and the same craving for sympathy, which has been the unprofitable fashion of this world in all ages. And no human being possessing such constitutionals has a better chance of being made unhappy by them than the backward, uninteresting, wrong-doing child. We can all sympathize, to some extent, with *men* and *women ;* but how few can go back to the sympathies of childhood; can understand the desolate insignificance of not being one of the *grown-up* people; of being sent to bed, to be *out of the way* in the evening, and to school, to be out of the way in the morning; of manifold similar grievances and distresses, which the child has no elocution to set forth, and the grown person no imagination to conceive.

A child's power. Ah, these children, little witches, pretty even in all their faults and absurdities. See, for example, yonder little fellow in a naughty fit. He has shaken his long curls over his deep-blue eyes, the fair brow is bent in a frown, the rose-leaf lip is pursed up in infinite defiance, and the white shoulder thrust angrily forward. Can any but a child look so pretty, even in its naughtiness ?

Then comes the instant change ; flashing smiles and tears, as the good comes back all in a rush, and you are overwhelmed with protestations, promises, and kisses! They are irresistible, too, these little ones. They pull away the scholar's pen, tumble about his paper, make somersets over his books; and what can he do? They tear up newspapers, litter the carpets, break, pull, and upset, and then jabber unheard-of English in self-defense; and what can you do for yourself?

The child as teacher. Wouldst thou know, O parent, what is that *faith* which unlocks heaven? Go not to wrangling polemics, or creeds and forms of theology, but draw to thy bosom thy little one, and read in that clear, trusting eye the lesson of eternal life. Be only to thy God as thy child is to thee, and all is done. Blessed shalt thou be, indeed, when " a little child shall lead thee."

PINK AND WHITE TYRANNY.

Baby's dreams. " An' it 's a blessin' they brings wid 'em to a house, sir; the angels come down wid 'em. We can't see 'em, sir; but, bless the darlin', she can. An' she smiles in her sleep when she sees 'em."

BETTY'S BRIGHT IDEA.

Mother pride. A heavenly amusement, such as that with which mothers listen to the foolish-wise prattle of children just learning to talk.

AGNES OF SORRENTO.

A child's defense. "The fact is, when I begin to talk, she gets her arms around my old neck, and falls to weeping and kissing me at such a rate as makes a fool of me. If the child would only be rebellious, one could do something; but this love takes all the stiffness out of one's joints."

UNCLE TOM'S CABIN.

Child's mission. "What would the poor and lowly do without children?" said St. Clare, leaning on the railing, and watching Eva, as she tripped off, leading Tom with her. "Your little child is your only true Democrat. Tom, now, is a hero to Eva; his stories are wonders in her eyes, his songs and Methodist hymns are better than an opera, and the traps and little bits of trash in his pocket a mine of jewels, and he the most wonderful Tom that ever wore a black skin. This is one of the roses of Eden, that the Lord has dropped down expressly for the poor and lowly, who get few enough of any other kind."

Animation. She was one of those busy, tripping creatures, that can no more be contained in one place than a sunbeam or a summer breeze.

SUNNY MEMORIES OF FOREIGN LANDS.

Unperverted taste. Children are unsophisticated, and like sugar better than silver any day.

POGANUC PEOPLE.

Child-faith. Dolly was at the happy age when anything bright and heavenly seemed credible, and had the child-faith to which all things are possible. She had even seriously pondered, at times, the feasibility of walking some day to the end of the rainbow, to look for the pot of gold which Nabby had credibly assured her was to be found there; and if at any time in her ramblings through the woods a wolf had met her, and opened a conversation, as in the case of Little Red Riding Hood, she would have been no way surprised, but kept up her part of the interview with becoming spirit.

LITTLE PUSSY WILLOW.

Simplicity. " Mother," she said, soberly, when she lay down in her little bed that night, "I'm going to ask God to keep me humble."

" Why, my dear ? "

" Because I feel tempted to be proud, — I can make such good bread ! "

A DOG'S MISSION.

Hobbies. He bores everybody to death with his locomotive as artlessly as grown people do with their hobbies.

Our Charley. When the blaze of the wood-fire flickers up and down in our snug evening parlor, there dances upon the wall a little shadow, with a pug nose, a domestic household shadow — a busy shadow — a little restless specimen of perpetual motion, and the owner thereof is "our Charley." Now we should not write about him and his ways, if he were strictly a peculiar and individual existence of our own home-circle; but it is not so. "Our Charley" exists in a thousand, nay, a million families; he has existed in millions in all time back; his name is variously rendered in all the tongues of the earth; in short, we take "our Charley" in a generic sense, and we mean to treat of him as a little copy of the grown man — enacting in a shadowy ballet by the fireside all that men act in earnest in after life. He is a looking-glass for grown people, in which they may see how certain things become them — in which they may sometimes even see streaks and gleamings of something wiser than all the harsh conflict of life teaches them.

MY WIFE AND I.

Heavenly children. It seems to me that lovely and loving childhood, with its truthfulness, its frank sincerity, its pure, simple love, is so sweet and holy an estate that it would be a beautiful thing in heaven to have a band of heavenly children, guileless, gay, and forever joyous — tender spring blossoms of the Kingdom of Light. Was it of such whom He had left in his heavenly home our Savior was thinking, when He took little children up in his arms, and blessed them, and said, "of such is the Kingdom of Heaven?"

Poetry and prose. The first child in a family is its poem, — it is a sort of nativity play, and we bend before the young stranger with gifts, "gold, frankincense, and myrrh." But the tenth child in a poor family is *prose*, and gets simply what is due to comfort. There are no superfluities, no fripperies, no idealities, about the tenth cradle.

A child's crosses. My individual pursuits, my own little stock of interests, were of course of no account. I was required to be in a perfectly free, disengaged state of mind, and ready to drop everything at a moment's warning from any of my half-dozen seniors. "Here, Hal, run down cellar and get me a dozen apples," my brother would say, just as I had half-built a block-house. "Harry, run upstairs and get the book I left on the bed " — " Harry, run out to the barn and get the rake I left there " — " Here, Harry, carry this up garret " — " Harry, run out to the tool-shop and get that " — were sounds constantly occurring — breaking up my private, cherished little enterprise of building cob-houses, making mill-dams and bridges, or loading carriages, or driving horses. Where is the mature Christian who could bear with patience the interruptions and crosses in his daily schemes that beset a boy?

Repression. When children grow up among older people, and are pushed and jostled and set aside in the more engrossing interests of their elders, there is an almost incredible amount of timidity and dumbness of nature, with regard to the expression of inward feeling, — and yet, often at this time, the instinctive sense of pleasure and pain is fearfully acute. But the child has imperfectly learned language ; his stock of words, as yet, consists only in names and attributes of outward and physical objects, and he has no phraseology with which to embody a mere emotional experience.

CHAPTER V.

EDUCATION.

THE MINISTER'S WOOING.

Habit. A man cannot ravel out the stitches in which early days have knit him.

Human error. All systems that deal with the infinite are, besides, exposed to danger from small, unsuspected admixtures of human error, which become deadly when carried to such vast results. The smallest speck of earth's dust, in the focus of an infinite lens, appears magnified among the heavenly orbs as a frightful monster.

Defective education. True it is, that one can scarcely call *that* education which teaches woman everything except herself, — except the things that relate to her own peculiar womanly destiny, and, on plea of the holiness of ignorance, sends her without a word of just counsel into the temptations of life.

OLDTOWN FOLKS.

Education of man and woman. The problem of education is seriously complicated by the peculiarities of womanhood. If we suppose two souls, exactly alike, sent into bodies, the one of man, the other of woman, that mere fact alone alters the whole mental and moral history of the two.

SAM LAWSON'S STORIES.

"Keep straight on." "Wal, ye see, boys, that 'ere 's jest the way to fight the Devil. Jest keep straight on with what ye 're doin', an' don't ye mind him, an' he can't do nothin' to ye."

Letting go. "Lordy massy! what can any on us do? There 's places where folks jest lets go 'cause they hes to. Things ain't as they want 'em, an' they can't alter 'em."

PEARL OF ORR'S ISLAND.

A mutual education. Those who contend against giving woman the same education as man do it on the ground that it would make the woman unfeminine, — as if Nature had done her work so slightly that it could be so easily raveled and

knit over. In fact, there is a masculine and feminine element in all knowledge, and a man and a
woman, put to the same study, extract only what
their nature fits them to see — so that knowledge
can be fully orbed only when the two unite in
the search and share the spoils.

Baiting the   "But don't you think Moses shows some
boy.   taste for reading and study?"

"Pretty well, pretty well!" said Zephaniah.
"Jist keep him a little hungry, not let him get all
he wants, you see, and he 'll bite the sharper.
If I want to catch cod I don't begin with flingin'
over a barrel o' bait. So with the boys, jist bait
'em with a book here an' a book there, an' kind
o' let 'em feel their own way, an' then, if nothin'
will do but a feller must go to college, give in to
him, — that 'd be my way."

A natural   "Colleges is well enough for your
education. smooth, straight-grained lumber, for
gen'ral buildin'; but come to fellers that 's got
knots an' streaks, an' cross-grains, like Moses
Pennel, an' the best way is to let 'em eddicate
'emselves, as he 's a-doin.' He 's cut out for the
sea, plain enough, an' he 'd better be up to Umba-
gog, cuttin' timber for his ship, than havin' rows
with tutors, an' blowin' the roof off the colleges,
as one o' them 'ere kind o' fellers is apt to, when
he don't have work to use up his steam. Why,
mother, there 's more gas got up in them Bruns-
wick buildin's from young men that are spilin'
for hard work than you could shake a stick at."

LITTLE FOXES.

Recreation. The true manner of judging of the worth of amusements is to try them by their effects on the nerves and spirits the day after. True amusement ought to be, as the word indicates, *recreation*, — something that refreshes, turns us out anew, rests the mind and body by change, and gives cheerfulness and alacrity to our return to duty.

Making the best of it. The principal of a large and complicated public institution was complimented on maintaining such uniformity of cheerfulness amid such a diversity of cares. " I 've made up my mind to be satisfied, when things are done half as well as I would have them," was his answer, and the same philosophy would apply with cheering results to the domestic sphere.

Individuality. Every human being has some handle by which he may be lifted, some groove in which he was meant to run ; and the great work of life, as far as our relations with each other are concerned, is to lift each one by his own proper handle, and run each one in his own proper groove.

HOUSE AND HOME PAPERS.

Need of home attractions. Parents may depend upon it that, if they do not make an attractive resort for their boys, Satan will. There are places enough, kept warm and light, and bright and merry, where boys can go whose mothers' parlors are too fine for them to sit in. There are enough to be found to clap them on the back, and tell them stories that their mothers must not hear, and laugh when they compass with their little piping voices the dreadful litanies of sin and shame.

Home education. The word home has in it the elements of love, rest, permanency, and liberty; but besides these it has in it the idea of an education by which all that is purest within us is developed into nobler forms, fit for a higher life. The little child by the home fireside was taken on the Master's knee when he would explain to his disciples the mysteries of the kingdom.

The education of the parent. Education is the highest object of home, but education in the widest sense — education of the parents no less than of the children. In a true home, the man and the woman receive, through their cares, their watchings, their hospitality, their charity, the last and highest finish that earth can put upon them. From that they must pass upward, for earth can teach them no more.

Perfection in little thiugs. To do common things perfectly is far better worth our endeavor than to do uncommon things respectably.

The cross. Right on the threshold of all perfection lies the *cross* to be taken up. No one can go over or around that cross in science or in art. Without labor and self-denial neither Raphael nor Michael Angelo nor Newton was made perfect.

THE CHIMNEY CORNER.

A well-developed man. We still incline to class distinctions and aristocracies. We incline to the scheme of dividing the world's work into two classes: first, physical labor, which is held to be rude and vulgar, and the province of a lower class; and second, brain-labor, held to be refined and aristocratic, and the province of a higher class. Meanwhile the Creator, who is the greatest of levelers, has given to every human being *both* a physical system, needing to be kept in order by physical labor, and an intellectual or brain power, needing to be kept in order by brain labor. *Work,* use, employment, is the condition of health in both; and he who works either to the neglect of the other lives but a half-life, and is an imperfect human being.

THE MAYFLOWER.

Intemperance. It is a great mistake to call nothing intemperance but that degree of physical excitement which completely overthrows the mental powers. There is a state of nervous excitability, resulting from what is often called moderate stimulation, which often long precedes this, and is, in regard to it, like the premonitory warnings of the fatal cholera — an unsuspected draft on the vital powers, from which, at any moment, they may sink into irremediable collapse.

It is in this state, often, that the spirit of gambling or of wild speculation is induced by the morbid cravings of an over-stimulated system. Unsatisfied with the healthy and regular routine of business, and the laws of gradual and solid prosperity, the excited and unsteady imagination leads its subjects to daring risks, with the alternative of unbounded gain on the one side, or of utter ruin on the other. And when, as is too often the case, that ruin comes, unrestrained and desperate intemperance is the wretched resort to allay the ravings of disappointment and despair.

Religious instruction at home. The only difficulty, after all, is that the keeping of the Sabbath and the imparting of religious instruction are not made enough of a *home* object. Parents pass off the responsibility on to the Sunday-school teacher, and suppose, of course, if they send their children to Sunday-school, they do the best they can for them. Now, I am satisfied, from my experience as a Sabbath-school teacher, that the best religious instruction imparted abroad still stands in need of the coöperation of a systematic plan of religious discipline and instruction at home ; for, after all, God gives a power to the efforts of a *parent* that can never be transferred to other hands.

What girls should be taught. If, amid the multiplied schools, whose advertisements now throng our papers, purporting to teach girls everything, both ancient and modern, high and low, from playing on the harp and working pin-cushions up to civil engineering, surveying, and navigation, there were any which could teach them to be women, — to have thoughts, opinions, and modes of action of their own, — such a school would be worth having. If one half of the good purposes which are in the hearts of the ladies of our nation were only acted out without fear of anybody's opinion, we should certainly be a step nearer the millennium.

PINK AND WHITE TYRANNY.

Dangers of vanity. She had the misfortune — and a great one it is — to have been singularly beautiful from the cradle, and so was praised and

exclaimed over and caressed as she walked the
streets. She was sent for far and near; borrowed
to be looked at; her picture taken by photogra-
phers. If one reflects how many foolish and
inconsiderate people there are in the world, who
have no scruple in making a pet and plaything of
a pretty child, one will see how this one unlucky
lot of being beautiful in childhood spoiled Lillie's
chances of an average share of good sense and
goodness. The only hope for such a case lies in
the chance of possessing judicious parents.

AGNES OF SORRENTO.

Patient waiting. "Gently, my son! gently!" said the
monk; "nothing is lost by patience.
See how long it takes the good Lord to make a
fair flower out of a little seed; and He does all
quietly, without bluster. Wait on Him a little
in peacefulness and prayer, and see what He will
do for thee."

UNCLE TOM'S CABIN.

"Bobser-vation." "Well, yer see," said Sam, proceeding
gravely to wash down Haley's pony,
"I 'se 'quired what ye may call a habit o' *bobser-
vation*, Andy. It 's a very 'portant habit, Andy,
and I 'commend yer to be cultivatin' it, now
ye 'r' young. Hist up that hind foot, Andy.

Yer see, Andy, it 's bobservation makes all der difference in niggers. Did n't I see which way de wind blew dis yer mornin'? Did n't I see what missis wanted, though she never let on? Dat ar' 's bobservation, Andy. I 'spects it 's what you may call a faculty. Faculties is different in different peoples, but cultivatin' of 'em goes a great way."

Honoring mother. "Now, Mas'r George," said Tom, "ye must be a good boy; 'member how many hearts is sot on ye. Al'ays keep close to yer mother. Don't be gettin' into any o' them foolish ways boys has of gettin' too big to mind their mothers. Tell ye what, Mas'r George, the Lord gives good many things twice over; but he don't give ye a mother but once. Ye 'll never see sich another woman, Mas'r George, if ye live to be a hundred years old. So, now, you hold on to her, and grow up and be a comfort to her, thar 's my own good boy, — you will, now, won't ye?"

DRED.

Silent influence. "Nina, I know, will love you; and if you never *try* to advise her and influence her, you will influence her very much. Good people are a long while learning *that*, Anne. They think to do good to others by interfering and advising. They don't know that all they have to do is to live."

Starting
right.
It is only the first step that costs.

LITTLE PUSSY WILLOW.

The right
way to
study.
"Knowledge has just been rubbed on to me upon the outside, while you have opened your mind, and stretched out your arms to it, and taken it in with all your heart."

A DOG'S MISSION.

The turn-
ing-point
in life.
There is an age when the waves of manhood pour in on the boy like the tides in the Bay of Fundy. He does not know himself what to do with himself, and nobody else knows either ; and it is exactly at this point that many a fine fellow has been ruined for want of faith and patience and hope in those who have the care of him.

"What
shall we
do with
Charley?"
But, after all, Charley is not to be wholly shirked, for he is an institution, a solemn and awful *fact;* and on the answer of the question, What is to be done with him? depends a future. Many a hard, morose, and bitter man has come from a Charley turned off and neglected ; many a parental heartache has come from a Charley left to run the streets, that mamma and sisters might play on the piano and write letters in peace. It is easy to get rid of him — there are fifty ways of doing that —

he is a spirit that can be promptly laid for a season, but if not laid aright, will come back by and by a strong man armed, when you cannot send him off at pleasure.

Mamma and sisters had better pay a little tax to Charley now, than a terrible one by and by. There is something significant in the old English phrase, with which our Scriptures make us familiar, — a *man* child ! A *man* child ! There you have the word that should make you think more than twice before you answer the question, What shall we do with Charley ?

For to-day he is at your feet — to-day you can make him laugh; you can make him cry; you can persuade, and coax, and turn him to your pleasure ; you can make his eyes fill and his bosom swell with recitals of good and noble deeds ; in short, you can mold him, if you will take the trouble.

But look ahead some years, when that little voice shall ring in deep bass tones ; when that small foot shall have a man's weight and tramp ; when a rough beard shall cover that little round chin, and all the strength of manhood fill out that little form. Then, you would give worlds to have the key to his heart, to be able to turn and guide him to your will ; but if you lose that key now he is little, you may search for it carefully with tears some other day, and not find it. Old housekeepers have a proverb that one hour lost in the morning is never found all day — it has a significance in this case.

MY WIFE AND I.

Limit of responsibility. One part of the science of living is to learn just what our own responsibility is, and to let other people's alone.

Starved faculties. People don't realize what it is to starve faculties ; they understand physical starvation, but the slow fainting and dying of desires and capabilities for want of anything to feed upon, the withering of powers for want of exercise, is what they do not understand.

Idealizing our work. The chief evil of poverty is the crushing of ideality out of life, taking away its poetry and substituting hard prose ; — and this, with them, was impossible. My father loved the work he did as the artist loves his painting, and the sculptor his chisel. A man needs less money when he is doing only what he loves to do — what, in fact, he *must* do, — pay or no pay. . . . My mother, from her deep spiritual nature, was one soul with my father in his life-work. With the moral organization of a prophetess she stood nearer to heaven than he, and looking in told him what she saw, and he, holding her hand, felt the thrill of celestial electricity.

True greatness. "I want you to be a good man. A great many have tried to be great men, and failed, but nobody ever sincerely tried to be a *good* man and failed."

Lack of religious instruction. But I speak from experience when I say that the course of study in Christian America is so arranged that a boy, from the grammar school upward till he graduates, is so fully pressed and overloaded with all other studies that there is no probability that he will find the time or the inclination for such (religious) investigations.

Educating boys for husbands. In our days we have heard much said of the importance of training women to be wives. Is there not something to be said on the importance of training men to be husbands? Is the wide latitude of thought and reading and expression which has been accorded as a matter of course to the boy and the young man, the conventionally allowed familiarity with coarseness and indelicacy, a fair preparation to enable him to be the intimate companion of a pure woman? For how many ages has it been the doctrine that man and woman were to meet in marriage, the one crystal-pure, the other foul with the permitted garbage of all sorts of uncleansed literature and license?

If the man is to be the head of the woman, even as Christ is the head of the Church, should he not be her equal, at least, in purity?

Moral courage. The pain-giving power is a most necessary part of a well-organized human being. Nobody can ever do anything without the courage to be disagreeable at times.

Appreciating individuality. Who is appreciative and many-sided enough to guide the first efforts of genius just coming to consciousness ? How many could profitably have advised Hawthorne when his peculiar Rembrandt style was forming ? As a race, we Anglo-Saxons are so self-sphered that we lack the power to enter into the individuality of another mind, and give profitable advice for its direction.

Truth told by an enemy. The truth, bitterly told by an enemy with a vivid power of statement, is a tonic oftentimes too strong for one's powers of endurance.

WE AND OUR NEIGHBORS.

Immutability of Nature's laws. In some constitutions, with some hereditary predispositions, the indiscretions and ignorances of youth leave a fatal, irremediable injury. Though the sin be in the first place one of inexperience and ignorance, it is one that nature never forgives. The evil once done can never be undone ; no prayers, no entreaties, no resolutions, can change the consequences of violated law. The brain and nerve force once vitiated by poisonous stimulants become thereafter subtle tempters and traitors, forever lying in wait to deceive, and urging to ruin; and he who is saved is saved so as by fire.

Doing our *own* work. "There must be second fiddles in an orchestra, and it's fortunate that I have precisely the talent for playing one, and my doctrine is that the second fiddle *well* played is quite as good as the first. What would the first be without it?"

Courage. "Well, there's no way to get through the world but to keep doing, and to attack every emergency with courage."

Value of truth. "We've got to get truth as we can in this world, just as miners dig gold out of the mines, with all the quartz, and dirt, and dross; but it pays."

CHAPTER VI.

NATURE.

THE MINISTER'S WOOING.

Want of sympathy in nature. The next day broke calm and fair. The robins sang remorselessly in the apple-tree, and were answered by bobolink, oriole, and a whole tribe of ignorant little bits of feathered happiness that danced among the leaves. Golden-glorious unclosed those purple eyelids of the east, and regally came up the sun; and the treacherous sea broke into a thousand smiles, laughing and dancing with every ripple, as unconsciously as if no form dear to human hearts had gone down beneath it. Oh, treacherous, deceiving beauty of outward things! beauty, wherein throbs not one answering nerve to human pain!

The sea. And ever and anon came on the still air the soft, eternal pulsations of the distant sea, — sound mournfullest, most mysterious, of all the harpings of nature. It was the sea, — the deep, eternal sea, — the treacherous, soft, dreadful, inexplicable sea.

OLDTOWN FOLKS.

The sunrise. The next morning showed as brilliant a getting-up of gold and purple as ever belonged to the toilet of a morning. There was to be seen from Asphyxia's bedroom window a brave sight, if there had been any eyes to enjoy it, — a range of rocky cliffs with little pin-feathers of black pine upon them, and behind them the sky all aflame with bars of massy light, — orange and crimson and burning gold, — and long bright rays, darting hither and thither, touched now the window of a farm-house, which seemed to kindle and flash back a morning salutation ; now they hit a tall, scarlet maple, now they pierced between clumps of pine, making their black edges flame with gold ; and over all, in the brightening sky, stood the morning star, like a great, tremulous tear of light, just ready to fall on the darkened world.

October in New England. Nature in New England is, for the most part, a sharp, determined matron of the Miss Asphyxia school. She is shrewd, keen, relentless, energetic. She runs through the seasons a merciless express-train, on which you may jump if you can, at her hours, but which knocks you down remorselessly if you come in her way, and leaves you hopelessly behind if you are late. Only for a few brief weeks in the autumn does this grim, belligerent female condescend to be charming; but when she does set about it, the veriest Circe of enchanted isles could not do it better. Airs more dreamy, more hazy, more full of purple light and lustre, never lay over Cyprus or Capri, than those which each October overshadow the granite rocks and prickly chestnuts of New England. The trees seem to run no longer sap, but some strange liquid glow; the colors of the flowers flame up, from the cold, pallid delicacy of spring, into royal tints wrought of the very fire of the sun, and the hues of evening clouds. The humblest weed, which we trod under our foot, unnoticed, in summer, changes with the first frost into some colored marvel, and lifts itself up into a study for a painter, just as the touch of death or adversity strikes out in a rough nature traits of nobleness and delicacy before wholly undreamed of.

THE CHIMNEY CORNER.

Gems. Gems, in fact, are a species of mineral flowers; they are the blossoms of the dark, hard mine; and what they want in perfume, they make up in durability.

THE MAYFLOWER.

Meditations of the oak. I sometimes think that leaves are the thoughts of trees, and that if we only

knew it, we should find their life's experience
recorded in them. Our oak — what a crop of
meditations and remembrances must he have
thrown forth, leafing out century after century!
Awhile he spake and thought only of red deer
and Indians; of the trillium that opened its
white triangle in his shade ; of the scented arbutus, fair as the pink ocean shell, weaving her
fragrant mats in the moss at his feet; of feathery
ferns, casting their silent shadows on the check-
erberry leaves, and all those sweet, wild, name-
less, half-mossy things that live in the gloom of
forests, and are only desecrated when brought to
scientific light, laid out, and stretched on a bo-
tanic bier. Sweet old forest days ! when blue
jay, and yellow-hammer, and bobolink made his
leaves merry, and summer was a long opera of
such music as Mozart dimly dreamed. But then
came human kind bustling beneath ; wondering, fussing, exploring, measuring, treading down
flowers, cutting down trees, scaring bobolinks,
and Andover, as men say, began to be settled.

The brook in winter. Let us stop the old chaise and get out a
minute to look at this brook, — one of
our last summer's pets. What is he doing this
winter ? Let us at least say "How do you do ?"
to him. Ah, here he is ! and he and Jack Frost
together have been turning the little gap in the
old stone wall, through which he leaped down
to the road, into a little grotto of Antiparos.
Some old rough rails and boards that dropped

over it are sheathed in plates of transparent
silver. The trunks of the black alders are
mailed with crystal; and the witch-hazel and
yellow osiers fringing its sedgy borders are like-
wise shining through their glossy covering.
Around every stem that rises from the water is
a glittering ring of ice. The tags of the alder
and the red berries of last summer's wild roses
glitter now like a lady's pendant. As for the
brook, he is wide-awake and joyful; and where
the roof of sheet ice breaks away, you can see his
yellow-brown waters rattling and gurgling among
the stones as briskly as they did last July. Down
he springs! over the glossy-coated stone wall,
throwing new sparkles into the fairy grotto
around him; and widening daily from melting
snows, and such other godsends, he goes chattering off under yonder mossy stone bridge, and we
lose sight of him. It might be fancy, but it
seemed that our watery friend tipped us a cheery
wink as he passed, saying, " Fine weather, sir
and madam; nice times these; and in April
you 'll find us all right; the flowers are making
up their finery for the next season; there 's to
be a splendid display in a month or two."

Trees in   Neither are trees, as seen in winter,
winter.   destitute of their own peculiar beauty.
If it be a gorgeous study in summer-time to
watch the play of their abundant foliage, we still
may thank winter for laying bare before us the
grand and beautiful anatomy of the tree, with all

its interlacing network of boughs, knotted on each twig with the buds of next year's promise. The fleecy and rosy clouds look all the more beautiful through the dark lace veil of yonder magnificent elms; and the down-drooping drapery of yonder willow hath its own grace of outline as it sweeps the bare snows. And the comical old apple-trees, why, in summer they look like so many plump, green cushions, one as much like another as possible; but under the revealing light of winter every characteristic twist and jerk stands disclosed.

One might moralize on this, — how affliction, which strips us of all ornaments and accessories, and brings us down to the permanent and solid wood of our nature, develops such wide differences in people who before seemed not much distinct.

Winter clouds. The cloud lights of a wintry sky have a clear purity and brilliancy that no other months can rival. The rose tints, and the shading of rose tint into gold, the flossy, filmy accumulation of illuminated vapor that drifts across the sky in a January afternoon, are beauties far exceeding those of summer.

AGNES OF SORRENTO.

Natural and moral elevation. There is always something of elevation and purity that seems to come over one from being in an elevated region. One feels morally as well as physically above the world, and from that clearer air able to look down on it calmly, with disengaged freedom.

The summit of Vesuvius. Around the foot of Vesuvius lie fair villages and villas garlanded with roses and flushing with grapes whose juice gains warmth from the breathing of its subterraneous fires, while just above them rises a region more awful than can be created by the action of any common causes of sterility. There, immense tracts sloping gradually upward show a desolation so peculiar, so utterly unlike every common solitude of nature, that one enters upon it with the shudder we give at that which is wholly unnatural. On all sides are gigantic serpent convolutions of black lava, their immense folds rolled into every conceivable contortion, as if, in their fiery agonies, they had struggled, and wreathed and knotted together, and then grown cold and black with the imperishable signs of those terrific convulsions upon them. Not a blade of grass, not a flower, not even the hardiest lichen, springs up to relieve the utter deathliness of the scene. The eye wanders from one black, shapeless mass to another, and there is ever the same suggestion of hideous monster life — of goblin convulsions and strange fiend-like agonies in some age gone by. One's very footsteps have an unnatural, metallic clink, and one's garments brushing over the rough surface are torn and fretted by its sharp, remorseless touch, as if its very nature were so pitiless and acrid that the slightest contact revealed it.

UNCLE TOM'S CABIN.

The morning star. Calmly the rosy hue of dawn was stealing into the room. The morning star stood, with its solemn, holy eye of light, looking down on the man of sin, from out the brightening sky. Oh, with what freshness, what solemnity and beauty, is each new day born; as if to say to insensate man, "Behold! thou hast one more chance! *Strive* for immortal glory!"

DRED.

A Southern thunder-shower. The day had been sultry, and it was now an hour or two past midnight, when a thunder-storm, which had long been gathering and muttering in the distant sky, began to develop its forces.

A low shivering sigh crept through the woods, and swayed in weird whistlings the tops of the pines; and sharp arrows of lightning came glittering down among the darkness of the branches, as if sent from the bow of some warlike angel. An army of heavy clouds swept in a moment across the moon; then came a broad, dazzling, blinding sheet of flame, concentrating itself on the top of a tall pine near where Dred was standing, and in a moment shivered all its branches to the ground as a child strips the leaves from a twig. . . .

The storm, which howled around him, bent the forest like a reed, and large trees, uprooted from the spongy and tremulous soil, fell crashing with a tremendous noise; but, as if he had been a dark spirit of the tempest, he shouted and exulted. . . .

Gradually the storm passed by; the big drops dashed less and less frequently; a softer breeze passed through the forest, with a patter like the clapping of a thousand little wings; and the moon occasionally looked over the silvery battlements of the great clouds.

Nature's lesson on love.   "Love is a mighty good ting, anyhow," said Tiff. "Lord bress you, Miss Nina, it makes eberyting go kind o' easy. Sometimes when I 'm studding upon dese yer tings, I says to myself, 'pears like de trees in de wood, dey loves each oder. Dey stands kind o' lockin' arms so, and dey kind o' nod der heads, and whispers so! 'Pears like de grapevines and de birds, and all dem ar tings, dey lives comfortable togeder, like dey was peaceable and liked each oder. Now, folks is apt to get a-stewin' an' a-frettin' round, an' turnin' up der noses at dis yer ting, an' dat ar; but 'pears like de Lord's works takes eberyting mighty easy. Dey jest kind o' lives along peaceable. I tink it 's mighty 'structive!"

PALMETTO LEAVES.

Winter, North and South. In New England, Nature is an up-and-down, smart, decisive house-mother, that has her times and seasons, and brings up her ends of life with a positive jerk. She will have no shilly-shally. When her time comes, she clears off the gardens and forests thoroughly and once for all, and they are clean. Then she freezes the ground solid as iron, and then she covers all up with a nice, pure winding-sheet of snow, and seals matters up as a good housewife does her jelly-tumblers under white paper covers. There you are, fast and cleanly. If you have not got ready for it, so much the worse for you! If your tender roots are not taken up, your cellar banked, your doors listed, she can't help it; it's your own lookout, not hers.

But Nature down here is an easy, demoralized, indulgent old grandmother, who has no particular time for anything, and does everything when she happens to feel like it. "Is it winter, or is n't it?" is the question likely often to occur in the settling month of December, when everybody up North has put away summer clothes, and put all their establishments under winter orders.

The olean-der. This bright morning we looked from the roof of our veranda, and our neighbor's oleander-trees were glowing like a great crimson cloud ; and we said, "There! the oleanders have come back!" No Northern ideas can give the glory of these trees as they raise their heads in this their native land, and seem to be covered with great crimson roses. The poor stunted bushes of Northern greenhouses are as much like it as our stunted virtues and poor, frost-nipped enjoyments shall be like the bloom and radiance of God's Paradise hereafter.

Moss. If you want to see a new and peculiar beauty, watch a golden sunset through a grove draperied with gray moss. The swaying, filmy bands turn golden and rose-colored, and the long, swaying avenues are like a scene in fairy-land.

The right side and the wrong. Every place, like a bit of tapestry, has its right side and its wrong side; and both are true and real, — the wrong side with its rags and tags, and seams and knots, and thrums of worsted, and the right side with its pretty picture.

SUNNY MEMORIES OF FOREIGN LANDS.

Beauty in nature. "Turn off my eyes from beholding vanity," says a good man, when he sees a display of graceful ornament. What, then, must he think of the Almighty Being, all whose useful work is so overlaid with ornament? There is not a fly's leg, not an insect's wing, which is not polished and decorated to an extent that we should think positive extravagance in finishing up a child's dress. And can we suppose that this Being can take delight in dwellings and modes of life and forms of worship where everything is reduced to cold, naked utility? I think not. The instinct to adorn and beautify is from Him; it likens us to Him, and if rightly understood, instead of being a siren to beguile our hearts away, it will be the closest affiliating band.

Flowers. There is a strange, unsatisfying pleasure about flowers, which, like all earthly pleasures, is akin to pain. What can you do with them? — you want to do something, but what? Take them all up and carry them with you? You cannot do that. Get down and look at them? What, keep a whole caravan waiting for your observation? That will never do. Well, then, pick and carry them along with you. That is what, in despair of any better resource, I did. . . . It seemed almost sacrilegious to tear away such fanciful creations, that looked as if they were votive offerings on an altar, or, more likely, living existences, whose only conscious life was a continual exhalation of joy and praise.

These flowers seemed to me to be the Earth's raptures and aspirations, — her better moments — her lucid intervals. Like everything else in our existence, they are mysterious.

In what mood of mind were they conceived by the great Artist? Of what feelings of His are

they the expression, — springing up out of the
dust, in the gigantic, waste, and desolate regions,
where one would think the sense of His almight-
iness might overpower the soul? Born in the
track of the glacier and the avalanche, they seem
to say to us that this Almighty Being is very
pitiful, and of tender compassion; that, in His
infinite soul, there is an exquisite gentleness and
love of the beautiful, and that, if we would be
blessed, His will to bless is infinite.

Mountain air. I look at the strange, old, cloudy mountains, the Eiger, the Wetterhorn, the
Schreckhorn. A kind of hazy ether floats around
them — an indescribable aerial halo — which no
painter ever represents. Who can paint the air,
— that vivid blue in which these sharp peaks cut
their glittering images?

The mys-
terious in
nature. I like best these snow-pure glaciers seen
through these black pines; there is
something mysterious about them when
you thus catch glimpses, and see not the earthly
base on which they rest. I recollect the same
fact in seeing the cataract of Niagara through
trees, where merely the dizzying fall of water
was visible, with its foam, and spray, and rainbow; it produced an idea of something supernatural. . . .
Every prospect loses by being made definite.
As long as we only see a thing by glimpses, and
imagine that there is a deal more that we do *not*

see, the mind is kept in a constant excitement
and play; but come to a point where you can
fairly and squarely take in the whole, and there
your mind falls listless. It is the greatest proof,
to me, of the infinite nature of our minds, that
we almost instantly undervalue what we have
thoroughly attained. . . . I remember once, after
finishing a very circumstantial treatise on the
nature of heaven, being oppressed with a similar
sensation of satiety, — that which hath not entered
the heart of man to conceive must not be mapped
out, — hence the wisdom of the dim, indefinite
imagery of the Scriptures; they give you no hard
outline, no definite limit; occasionally they part
as do the clouds around these mountains, giving
you flashes and gleams of something supernatural
and splendid, but never fully unveiling.

Cloud land- It is odd, though, to look at those cloud-
scapes.   caperings; quite as interesting, in its
way, as to read new systems of transcendental
philosophy, and perhaps quite as profitable.
Yonder is a great white-headed cloud, slowly
unrolling himself in the bosom of a black pine
forest. Across the other side of the road a huge
granite cliff has picked up a bit of gauzy silver,
which he is winding around his scraggy neck.
And now, here comes a cascade, right over our
heads; a cascade, not of water, but of cloud;
for the poor little brook that makes it faints away
before it gets down to us; it falls like a shimmer
of moonlight, or a shower of powdered silver,

while a tremulous rainbow appears at uncertain intervals, like a half-seen spirit.

A cascade. The cascade here, as in mountains generally, is a never-failing source of life and variety. Water, joyous, buoyant son of Nature, is calling to you, leaping, sparkling, mocking at you between bushes, and singing as he goes down the dells. A thousand little pictures he makes among the rocks as he goes.

Phases of nature. There are phases in nature which correspond to every phase of human thought and emotion; and this stern, cloudy scenery [in the Alps] answers to the melancholy fatalism of Greek tragedy, or the kindred mournfulness of the book of Job.

Sublimity in nature. Coming down I mentally compared Mont Blanc and Niagara, as one should compare two grand pictures in different styles of the same master. Both are of that class of things which mark eras in a mind's history, and open a new door which no man can shut. Of the two, I think Niagara is the more impressive, perhaps because those aerial elements of foam and spray give that vague and dreamy indefiniteness of outline which seems essential in the sublime. For this reason, while Niagara is equally impressive in the distance, it does not lose on the nearest approach, — it is always mysterious, and therefore stimulating. Those varying spray-wreaths, rising like Ossian's ghosts from its abyss; those shimmering rainbows, through whose veil you look; those dizzying falls of water that seem like clouds poured from the hollow of God's hand; and that mystic undertone of sound that seems to pervade the whole being as the voice of the Almighty, — all these bewilder and enchant the discriminating and prosaic part of us, and bring us into that cloudy region of ecstasy where the soul comes nearest to Him whom no eye has seen or can see. I have sometimes asked myself if, in the countless ages of the future, the heirs of God shall ever be endowed by Him with a creative power, by which they shall bring into being things like these? In this infancy of his existence, man creates pictures, statues, cathedrals; but when he is made "ruler over many things," will his Father intrust to him the building and adorning of worlds? the ruling of the glorious, dazzling forces of nature?

Mountain brooks. Everybody knows, even in our sober New England, that mountain brooks are a frisky, indiscreet set, rattling, chattering, and capering, in defiance of all law and order, tumbling over precipices and picking themselves up at the bottom, no whit wiser or more disposed to be tranquil than they were at the top; in fact, seeming to grow more mad and frolicsome with every leap. Well, that is just the way brooks do here in the Alps.

Alpine air. The whole air seemed to be surcharged with tints, ranging between the palest rose and the deepest violet — tints never without blue, and never without red, but varying in the degree of the two. It is this prismatic hue, diffused over every object, which gives one of the most noticeable characteristics of the Alpine landscape.

Color-blending. I have seen sometimes, in spring, set against a deep-blue sky, an array of greens, from lightest yellow to deepest blue of the pines, tipped and glittering with the afternoon sun, yet so swathed in some invisible, harmonizing medium, that the strong contrasts of color jarred upon no sense. All seemed to be bound by the invisible cestus of some celestial Venus. Yet what painter would dare attempt the same?

Nature's anguish. Mountains are nature's testimonials of anguish. They are the sharp cry of a groaning and travailing creation. Nature's stern agony writes itself on these furrowed brows of gloomy stone. These reft and splintered crags stand, the dreary images of patient sorrow, existing verdureless and stern because exist they must. In them, hearts that have ceased to rejoice, and have learned to suffer, find kindred, and here an earth worn with countless cycles of sorrow utters to the stars voices of speechless despair.

Pines. I always love pines, to all generations. I welcome this solemn old brotherhood, which stand gray-bearded, like monks, old, dark, solemn, sighing a certain mournful sound — like a *benedicite* through the leaves.

POGANUC PEOPLE.

New England spring. But at last — at last — spring did come at Poganuc! This marvel and mystery of the new creation *did* finally take place there every year, in spite of every appearance to the contrary. Long after the bluebird that had sung the first promise had gone back into his own celestial ether, the promise that he sang was fulfilled.

Like those sweet, foreseeing spirits, that on high, bare tree-tops of human thought, pour forth songs of hope in advance of their age and time, our bluebird was gifted with a sure spirit of prophecy; and, though the winds were angry and loud, though snows lay piled and deep for long weeks after, though ice and frost and hail armed themselves in embattled forces, yet the sun behind them all kept shining and shining, every day longer and longer, every day drawing nearer and nearer, till the snows passed away like a bad dream, and the brooks woke up and began to laugh and gurgle, and the ice went out of the ponds. Then the pussy-willows threw out their soft catkins, and the ferns came up with their woolly hoods on, like prudent old house-mothers, looking to see if it was yet time to unveil their tender greens, and the white blossoms of the shad-blow and the tremulous tags of the birches and alders shook themselves gayly out in the woods. Then, under brown, rustling leaf-banks, came the white, waxy shells of the trailing arbutus with its pink buds, fair as a winter's dawn on snow; the blue and white hepaticas opened their eyes, and cold, sweet, white violets starred the moist edges of water-courses, and great blue violets opened large eyes in the shadows, and the white and crimson trilliums unfurled under the flickering lace-work shadows of the yet leafless woods; the red columbine waved its bells from the rocks, and great tufts of golden cowslips fringed the borders of the brooks. Then came in flocks the delicate wind-flower family; anemones, starry white, and the crowfoot, with its pink outer shell, and the spotted adder's tongue, with its waving yellow bells of blossom. Then, too, the honest, great, green leaves of the old skunk-cabbage, most refreshing to the eye in its hardy, succulent greenness, though an abomination to the nose of the ill-informed who should be tempted to gather them. In a few weeks, the woods, late so frozen, — hopelessly buried in snow-drifts, — were full of a thousand beauties and delicacies of life and motion, and flowers bloomed on every hand.

Autumn. The bright days of summer were a short-lived joy at Poganuc. One hardly had time to say "How beautiful!" before it was past. By September came the frosty nights that turned the hills into rainbow colors, and ushered in Autumn, with her gorgeous robes of golden-rod and purple asters. There was still the best of sport for the children, however; for the frost ripened the shagbark walnuts and opened the chestnut burrs, and the glossy brown chestnuts dropped down among the rustling yellow leaves and the beds of fringed blue gentian. . . . Here and there groups of pines and tall hemlocks, with their heavy background of solemn green, threw out the flamboyant tracery of the forest in startling distinctness. Here and there, as they passed a bit of low land, the swamp maple seemed really to burn like crimson flame, and the clumps of black alder, with their vivid scarlet berries, exalted the effect of color to the very highest and most daring result. No artist ever has ventured to put on canvas the exact copy of the picture that Nature paints for us every year in the autumn months. There are things the Almighty Artist can do that no earthly imitator can more than hopelessly admire.

Bird-talk. Who shall interpret what is meant by the sweet jargon of robin and oriole and bobolink, with their endless reiterations? Something wiser, perhaps, than we dream of in our lower life here.

LITTLE PUSSY WILLOW.

New England winter. By and by the sun took to getting up later and later, setting a dreadfully bad example, it is to be confessed. It would be seven o'clock and after before he would show his red face above the bed-clothes of clouds away off in the southeast; and when he *did* manage to get up, he was so far off and so chilly in his demeanor that people seemed scarcely a bit the better for him; and by half-past four in the afternoon he was down in bed again, tucked up for the night, never caring what became of the world. And so the clouds were full of snow, as if a thousand white feather-beds had been ripped up over the world; and all the frisky winds came out of their dens, and great frolics they had, blowing and roaring and careering in the clouds, — now bellowing down between the mountains, as if they meant to tear the world to pieces, then piping high and shrill, first round one corner of the farm-house, and then round the other, rattling the windows, bouncing against the doors, and then with one united chorus rumbling, tumbling down the great chimney, as if they had a mind to upset it. Oh, what a frisky, rough, jolly, unmannerly set of winds they were! By and by the snow drifted higher than the fences, and nothing was to be seen around the farm-house but smooth, waving hills and hollows of snow; and then came the rain and sleet, and froze them over with a slippery, shining crust, that looked as if the earth was dressed for the winter in a silver coat of mail.

QUEER LITTLE PEOPLE.

Summer rain. There had been a patter of rain the night before, which had kept the leaves awake talking to each other till nearly morning, but by dawn the small winds had blown brisk little puffs, and had whisked the heavens clean and bright with their tiny wings, as you have seen Susan clear away the cobwebs in your mamma's parlor ; and so now there were left only a thousand blinking, burning water-drops, hanging like convex mirrors at the end of each leaf, and Miss Katy admired herself in each one.

MY WIFE AND I.

Influence of sur- roundings. The mutual acquaintance that comes to companions in this solitude and face-to- face communion with nature is deeper and more radical than can come when surround- ed by the factitious circumstances of society. When the whole artificial world is withdrawn, and far out of sight, when we are surrounded with the pure and beautiful mysteries of nature, the very best and most genuine part of us comes to the surface, we know each other by the com- munion of our very highest faculties.

RELIGIOUS POEMS.

Summer
studies.

Why shouldst thou study in the month
of June
In dusky books of Greek and Hebrew lore,
When the great Teacher of all glorious things
Passes in hourly light before thy door ?

There is a brighter book unrolling now ;
Fair are its leaves as is the tree of heaven,
All veined and dewed and gemmed with won-
    drous signs,
To which a healing, mystic power is given.

A thousand voices to its study call,
From the fair hill-top, from the water-fall,
Where the bird singeth, and the yellow bee,
And the breeze talketh from the airy tree.

Now is that glorious resurrection time
When all earth's buried beauties have new birth !
Behold the yearly miracle complete, —
God hath created a new heaven and earth !

Hast thou no *time* for all this wondrous show, —
No thought to spare ?   Wilt thou forever be
With thy last year's dry flower-stalk and dead
    leaves,
And no new shoot or blossom on thy tree ?

See how the pines push off their last year's leaves,
And stretch beyond them with exultant bound :

The grass and flowers, with living power, o'er-
    grow
Their last year's remnants on the greening
    ground.

Wilt thou, then, all thy wintry feelings keep,
The old dead routine of the book-writ lore,
Nor deem that God can teach, by one bright hour,
What life hath never taught to thee before ?

Cease, cease to *think*, and be content to *be ;*
Swing safe at anchor in fair Nature's bay;
Reason no more, but o'er thy quiet soul
Let God's sweet teachings ripple their soft way.

Call not such hours an idle waste of time, —
Land that lies fallow gains a quiet power;
It treasures, from the brooding of God's wings,
Strength to unfold the future tree and flower.

And when the summer's glorious show is past,
Its miracles no longer charm thy sight,
The treasured riches of those thoughtful hours
Shall make thy wintry musings warm and bright.

CHAPTER VII.

LITERATURE AND ART.

THE MINISTER'S WOOING.

Romance. All prosaic and all bitter, disenchanted people talk as if poets and novelists *made* romances. They do, just as much as craters make volcanoes, no more. What is romance? Whence comes it? Plato spoke to the subject wisely, in his quaint way, some two thousand years ago, when he said, "Man's soul, in a former state, was winged and soared among the gods: and so it comes to pass that, in this life, when the soul, by the power of music or poetry, or the sight of beauty, hath her remembrance quickened, forthwith there is a struggling and a pricking pain, as of wings trying to come forth, even as children in teething." And if an old heathen, two thousand years ago, discoursed thus gravely of the romantic part of our nature, whence comes it that in Christian lands we think in so pagan a way of it, and turn the whole care of it to ballad-makers, romancers, and opera-singers?

Let us look up in fear and reverence, and say, "God is the great maker of romance; He, from whose hand came man and woman, — He, who strung the great harp of existence, — He is the great poet of life." Every impulse of beauty, of heroism, and every craving for purer love, fairer perfection, nobler type and style of being, than that which closes like a prison-house around us, in the dim, daily walk of life, is God's breath, God's impulse, God's reminder to the soul that there is something higher, sweeter, purer, yet to be attained. . . .

The dullest street of the most prosaic town has matter in it for more smiles, more tears, more intense excitement, than ever were written in story or sung in poem; the reality is there, of which the romance is the second-hand recorder.

OLDTOWN FOLKS.

Hebrew literature. But it is a most remarkable property of this old Hebrew literature that it seems to be enchanted with a divine and living power, which strikes the nerve of individual consciousness in every desolate and suffering soul. It may have been Judah or Jerusalem ages ago to whom these words first came, but as they have traveled on for thousands of years, they have seemed to tens of thousands of sinking and desolate souls the voice of God to them individually. They have raised the burden from thousands of crushed spirits; they have been as the day-spring to thousands of perplexed wanderers. Oh! let us treasure these old words, for as of old Jehovah chose

to dwell in a tabernacle in the wilderness, and
between the cherubim in the temple, so now He
dwells in them; and to the simple soul that
seeks for Him here, He will look forth as of old
from the pillar of cloud and fire.

Influence
of the
Bible.
For my part, I am impatient of the
theory of those who think that nothing that is not understood makes any
valuable impression on the mind of a child. I
am certain that the constant contact of the Bible
with my childish mind was a very great mental
stimulant, as it certainly was a cause of a singular and vague pleasure. The wild, poetic parts
of the prophecies, with their bold figures, vivid
exclamations, and strange Oriental names and
images, filled me with a quaint and solemn delight. Just as a child brought up under the
shadow of the great cathedrals of the Old World,
wandering into them daily, at morning, or at
eventide, beholding the many-colored windows,
flamboyant with strange legends of saints and
angels, and neither understanding the legends,
nor comprehending the architecture, is yet stilled
and impressed, till the old minster grows into his
growth and fashions his nature, so this wonderful
old cathedral book insensibly wrought a sort of
mystical poetry into the otherwise hard and ster-
ile life of New England. Its passionate Oriental
phrases, its quaint, pathetic stories, its wild tran-
scendent bursts of imagery, fixed an indelible
mark in my imagination. . . . I think no New

Englander, brought up under the *régime* established by the Puritans, could really estimate how much of himself had actually been formed by this constant, face-to-face intimacy with Hebrew literature.

The study of a new language. I recommend everybody who wishes to try the waters of Lethe to study a new language, and learn to think in new forms ; it is like going out of one sphere of existence into another.

Greek. Greek is the morning land of languages, and has the freshness of early dew in it which will never exhale.

THE PEARL OF ORR'S ISLAND.

The Bible. "This 'ere old Bible, — why it's jest like yer mother — ye rove and ramble and cut up round the world without her a spell, and mebbe think the old woman ain't so fashionable as some ; but when sickness and sorrow comes, why there ain't nothin' else to go back to. Is there, now ? "

HOUSE AND HOME PAPERS.

Reading only for amusement. "But don't you think," said Marianne, "that there is danger in too much fiction ? "

"Yes," said I. "But the chief danger of all that class of reading is its *easiness*, and the indolent, careless mental habit it induces. A great deal of the reading of young people on all days is really reading to no purpose, its object being merely present amusement. It is a listless yielding of the mind to be washed over by a stream which leaves no fertilizing properties, and carries away by constant wear the good soil of thought. I should try to establish a barrier against this kind of reading, not only on Sunday, but on Monday, on Tuesday, and on all days. Instead, therefore, of objecting to any particular class of books for Sunday reading, I should say in general that reading merely for pastime, without any moral aim, is the thing to be guarded against. That which inspires no thought, no purpose, which steals away all our strength and energy, and makes the Sabbath a day of dreams, is the reading I would object to."

Sacred music. "So of music. I do not see the propriety of confining one's self to technical sacred music. Any grave, solemn, thoughtful, or pathetic music has a proper relation to our higher spiritual nature, whether it be printed in a church service-book or on secular sheets. On me, for example, Beethoven's Sonatas have a far more deeply religious influence than much that has religious names and words. Music is to be judged of by its effects."

A good picture. For a picture, painted by a real artist, who studies Nature minutely and conscientiously, has something of the charm of the good Mother herself, — something of her faculty of putting on different aspects under different lights.

PINK AND WHITE TYRANNY.

Letters. Those long letters in which thoughtful people who live in retired situations delight ; letters, not of outward events, but of sentiments and opinions, the phases of the inner life.

AGNES OF SORRENTO.

The artist's mission. What higher honor or grace can befall a creature than to be called upon to make visible to men that beauty of invisible things which is divine and eternal ?

Hymns. "A hymn is a singing angel, and goes walking through the earth, scattering the devils before it. Therefore he who creates hymns imitates the most excellent and lovely works of our Lord God, who made the angels. These hymns watch our chamber-door, they sit upon our pillow, they sing to us when we awake ; and therefore our master was resolved to sow the minds of his young people with them, as our lovely Italy is sown with the seeds of all-colored flowers."

Music the language of Italy. There is no phase of the Italian mind that has not found expression in its music.

DRED.

The universal book. As the mind, looking on the great volume of nature, sees there a reflection of its own internal passions, and seizes on that in it which sympathizes with itself, — as the fierce and savage soul delights in the roar of torrents, the thunder of avalanches, and the whirl of ocean-storms, — so is it in the great answering volume of revelation. There is something there for every phase of man's nature, and hence its endless vitality and stimulating force.

Prophecy and Revelation. It is remarkable that in all ages, communities and individuals who have suffered under oppression have always fled for refuge to the Old Testament and to the book of Revelation in the New. Even if not definitely understood, these magnificent compositions have a wild, inspiring power, like a wordless, yet impassioned symphony, played by a sublime orchestra, in which deep and awful sub-bass instruments mingle with those of ethereal softness, and wild minors twine and interlace with marches of battles and bursts of victorious harmony.

They are much mistaken who say that nothing is efficient as a motive that is not definitely understood. Who ever thought of understanding the mingled wail and roar of the Marseillaise? Just this kind of indefinite stimulating power has the Bible to the souls of the oppressed. There is also a disposition, which has manifested itself since the primitive times, by which the human soul, bowed down beneath the weight of mighty oppressions, and despairing in its own weakness, seizes with avidity the intimations of a coming judgment, in which the Son of Man, appearing in His glory, and all His holy angels with Him, shall right earth's mighty wrongs.

SUNNY MEMORIES OF FOREIGN LANDS.

The artist as prophet. But, I take it, every true painter, poet, and artist is in some sense so far a prophet that his utterances convey more to other minds than he himself knows; so that, doubtless, should all the old masters rise from the dead, they might be edified by what posterity has found in their books.

Difficulty of criticism. Certainly no emotions so rigidly reject critical restraint and disdain to be bound by rule as those excited by the fine arts. A man unimpressible and incapable of moods and tenses is for that reason an incompetent critic; and the sensitive, excitable man, how can he know that he does not impose his peculiar mood as a general rule ?

Rembrandt and Hawthorne. I always did admire the gorgeous and solemn mysteries of his coloring. Rembrandt is like Hawthorne. He chooses simple and every-day objects, and so arranges light and shadow as to give them a sombre richness and mysterious gloom. "The House of the Seven Gables" is a succession of Rembrandt pictures, done in words instead of oils. Now, this pleases us, because our life really is a haunted one, the simplest thing in it *is* a mystery, the invisible world always lies around us like a shadow, and therefore this dreamy, golden gleam of Rembrandt meets somewhat in our inner consciousness to which it corresponds. . . .

Rubens and Shakespeare. I should compare Rubens to Shakespeare, for the wonderful variety and vital force of his artistic power. I know no other mind he so nearly resembles. Like Shakespeare, he forces you to accept and to forgive a thousand excesses, and uses his own faults as musicians use discords, only to enhance the perfection of harmony. There certainly is some use, even in defects. A faultless style sends you to sleep. Defects rouse and excite the sensibility to seek and appreciate excellence. Some of Shakespeare's finest passages explode all grammar and rhetoric like sky-rockets — the thought blows the language to shivers.

Language of the Bible. I rejoice every hour that I am among these scenes in my familiarity with the language of the Bible. In it alone can I find vocabulary and images to express what this world of wonder excites.

The effect of Christianity. As to Christianity not making men happier, methinks M. Belloc forgets that the old Greek tragedies are filled with despair and gloom, as their prevailing characteristic, and that nearly all the music of the world before Christ was in the minor scale, as since Christ it has come to be in the major. The whole creation has, indeed, groaned and travailed in pain together until now, but the mighty anthem has modulated since the Cross, and the requiem of Jesus has been the world's birth-song of approaching jubilee.

Music is a far better test, moreover, on such a point, than painting, for just where painting is weakest, namely, in the expression of the highest moral and spiritual ideas, there music is most sublimely strong.

Real music. To me, all music is sacred. Is it not so? All *real* music, in its passionate earnest, its blendings, its wild, heart-searching tones, is the language of aspiration. So it may not be meant; yet, when we know God, so we translate it.

Power of inward emotion. What is done from a genuine, strong, inward emotion, whether in writing or painting, always mesmerizes the paper or the canvas, and gives it a power which everybody must feel, though few know why. The reason why the Bible has been omnipotent, in all ages, has been because there were the emotions of God in it.

POGANUC PEOPLE.

Puritan music. As there is a place for all things in this great world of ours, so there was in its time and day a place and a style for Puritan music. If there were pathos and power and solemn splendor in the rhythmic movement of the churchly chants, there was a grand, wild freedom and energy of motion in the old " fuguing tunes " of that day that well expressed the heart of a people courageous in combat and unshaken in endurance. The church chant is like the measured motion of the mighty sea in calm weather, but those old fuguing tunes were like that same ocean aroused by stormy winds, when deep calleth unto deep in tempestuous confusion, out of which, at last, is evolved union and harmony. It was a music suggestive of the strife, the commotion, the battle-cries of a transition period of society, struggling onward toward dimly seen ideals of peace and order.

LITTLE PUSSY WILLOW.

Books. No ornament of a house can compare with books; they are constant company in a room, even when you are not reading them.

MY WIFE AND I.

Our thoughts in others' words. The only drawback when one reads poems that exactly express what one would like to say is that it makes us envious; one thinks, why could n't I have said it thus?

WE AND OUR NEIGHBORS.

Books of meditation. St. John was seated in his study, with a book of meditations before him, on which he was endeavoring to fix his mind. In the hot, dusty, vulgar atmosphere of modern life, it was his daily effort to bring around himself the shady coolness, the calm, conventual stillness, that breathes through such writers as St. Francis de Sales and Thomas à Kempis, men with a genius for devotion, who have left to mankind records of the milestones and road-marks by which they traveled towards the highest things. Nor should the most stringent Protestant fail to honor that rich and grand treasury of the experience of devout spirits of which the Romish

Church has been the custodian. The hymns and prayers and pious meditations which come to us through this channel are particularly worthy of a cherishing remembrance in this dusty, materialistic age.

Hymns. Words of piety, allied to a catching tune, are like seeds with wings — they float out in the air, and drop in the odd corners of the heart, to spring up in good purposes.

CHAPTER VIII.

NEW ENGLAND LIFE.

THE MINISTER'S WOOING.

Earnestness of the New England people. It is impossible to write a story of New England life and manners for a thoughtless, shallow-minded person. If we represent things as they are, their intensity, their depth, their unworldly gravity and earnestness must inevitably repel lighter spirits, as the reverse pole of the magnet drives off sticks and straws. In no other country were the soul and the spiritual life ever such intense realities, and everything contemplated so much (to use a current New England phrase) " in reference to eternity."

New England theology. The rigid theological discipline of New England is fitted to produce rather strength and purity than enjoyment. It was not fitted to make a sensitive and thoughtful nature happy, however it might ennoble and exalt.

The kitchen. The kitchen of a New England matron was her throne-room, her pride ; it was

the habit of her life to produce the greatest possible results there with the slightest possible dis-
composure; and what any woman could do, Mrs.
Katy Scudder could do *par excellence*. Everything there seemed to be always done and never
doing. Washing and baking, those formidable
disturbers of the composure of families, were all
over within those two or three morning hours
when we are composing ourselves for a last nap,
— and only the fluttering of linen over the green
yard on Monday mornings proclaimed that the
dreaded solemnity of a wash had transpired. A
breakfast arose there as by magic; and in an
incredibly short space after, every knife, fork,
spoon, and trencher, clean and shining, was looking as innocent and unconscious in its place as
if it never had been used and never expected
to be.

The floor, — perhaps, sir, you remember your
grandmother's floor of snowy boards sanded with
whitest sand; you remember the ancient fireplace stretching quite across one end, — a vast
cavern, in each corner of which a cozy seat
might be found, distant enough to enjoy the
crackle of the great jolly wood fire; across the
room ran a dresser, on which was displayed
great store of shining pewter dishes and platters,
which always shone with the same mysterious
brightness; by the side of the fire, a commodi-
ous wooden settee, or "settle," offered repose to
people too little accustomed to luxury to ask for
a cushion. Oh, that kitchen of the olden time,

— the old, clean, roomy, New England kitchen! Who that has breakfasted, dined, and supped in one, has not cheery visions of its thrift, its warmth, its coolness? The noonmark on its floor was a dial that told off some of the happiest days; thereby did we right up some of the shortcomings of the solemn old clock that tick-tacked in the corner, and whose ticks seemed mysterious prophecies of unknown good yet to arise out of the hours of life. How dreamy the winter twilight came in there, — when as yet the candles were not lighted, — when the crickets chirped around the dark stone hearth, and shifting tongues of flame flickered and cast dancing shadows and elfish lights on the walls, while grandmother nodded over her knitting-work, and puss purred, and old Rover lay dreamily opening now one eye and then the other on the family group! With all our ceiled houses, let us not forget our grandmother's kitchen.

Faculty. She was one of the much admired class, who, in the speech of New England, are said to have faculty, a gift which, among that shrewd people, commands more esteem than beauty, riches, learning, or any other worldly endowment. *Faculty* is Yankee for *savoir faire*, and the opposite virtue to shiftlessness. Faculty is the greatest virtue, and shiftlessness the greatest vice, of Yankee men and women. To her who has faculty nothing shall be impossible. She shall scrub floors, wash, wring, bake, brew, and yet her hands shall be small and white; she shall have no perceptible income, yet always be handsomely dressed; she shall have not a servant in her house, — with a dairy to manage, hired men to feed, a boarder or two to care for, unheard-of pickling and preserving to do, — and yet you commonly see her every afternoon sitting at her shady parlor-window behind the lilacs, cool and easy, hemming muslin cap-strings, or reading the last new book. She who hath faculty is never in a hurry, never behindhand. She can always step over to distressed Mrs. Smith, whose jelly won't come, — and stop to show Mrs. Jones how she makes her pickles so green, — and be ready to watch with poor old Mrs. Simpkins, who is down with the rheumatism.

Garrets. Garrets are delicious places in any case, for people of thoughtful, imaginative temperament. Who has not loved a garret in the twilight days of childhood, with its endless stores of quaint, cast-off, suggestive antiquity, — old, worm-eaten chests, — rickety chairs, — boxes and casks full of odd comminglings, out of which, with tiny, childish hands, we picked wonderful hoards of fairy treasure? What peep-holes, and hiding-places, and undiscoverable retreats we made to ourselves, — where we sat rejoicing in our security, and bidding defiance to the vague, distant cry which summoned us to school, or to some unsavory every-day task! How deliciously the rain came pattering on the roof over our head, or the red twilight streamed in at the window, while we sat snugly ensconced over the delicious pages of some romance which careful aunts had packed away at the bottom of all things, to be sure we should never read it! If you have anything, beloved friends, which you wish your Charley or your Susy to be sure and read, pack it mysteriously away at the bottom of a trunk of stimulating rubbish in the darkest corner of your garret; in that case, if the book be at all readable, — one that by any possible chance can make its way into a young mind, you may be sure that it will not only be read, but remembered to the longest day they have to live.

Clearcut thought. His was one of those clearly cut minds which New England forms among her farmers, as she forms quartz crystals in her mountains, by a sort of gradual influence flowing through every pore of the soil and system.

OLDTOWN FOLKS.

New England the parent of the West. New England has been to the United States what the Dorian hive was to Greece. It has always been a capital country to emigrate from, and North, South, East, and West have been populated largely from New England, so that the seed-bed of New England was the seed-bed of the great American republic, and of all that is likely to come of it.

Rough exterior. Any one that has ever pricked his fingers in trying to force open a chestnut-burr may perhaps have moralized at the satin lining, so smooth and soft, that lies inside of that sharpness. It is an emblem of a kind of nature very frequent in New England, where the best and kindest and most desirable of traits are enveloped in an outside wrapping of sharp austerity.

The do-nothing. Every New England village, if you only think of it, must have its do-nothing, as regularly as it has its school-house or its meeting-house. Nature is always wide awake in the matter of compensation. Work, thrift, and industry are such an incessant steam-power in Yankee life that society would burn itself out with the intense friction, were there not interposed here and there the lubricating power of a decided do-nothing, — a man who won't be hurried, and won't work, and will take his ease in his own way, in spite of the whole protest of his neighborhood to the contrary. And there is on the face of the whole earth no do-nothing whose softness, idleness, general inaptitude to labor, and everlasting, universal shiftlessness, can compare with that of the worthy, as found in a brisk Yankee village.

Life an engrossing interest. People have often supposed, because the Puritans founded a society where there were no professed public amusements, that therefore there was no fun going on in the ancient land of Israel, and that there were no cakes and ale, because they were virtuous. They were never more mistaken in their lives. There was an abundance of sober, well-considered merriment, and the hinges of life were well-oiled with that sort of secret humor which to this day gives the raciness to real Yankee wit. Besides this, we must remember that life itself is the greatest possible amusement to people who really believe they can do much with it, — who have that intense sense of what can be brought to pass by human effort that was characteristic of the New England colonies. To such, it is not exactly proper to say that life is an amusement, but it certainly is an engrossing interest, that takes the place of all amusements.

New England nobility. In the little theocracy which the Pilgrims established in the wilderness, the ministry was the only order of nobility. They were the only privileged class, and their voice it was that decided *ex cathedra* on all questions both of church and state, from the choice of governor to that of district school teacher.

Our minister, as I remember him, was one of the cleanest, most gentlemanly, most well-bred of men, — never appearing without all the decorums of silk stockings, shining knee and shoe buckles, well-brushed shoes, immaculately powdered wig, out of which shone his clear, calm, serious face, like the moon out of a fleecy cloud.

THE PEARL OF ORR'S ISLAND.

A ship-building community.

In the plain, simple regions we are describing, — where the sea is the great avenue of active life, and the pine forests are the great source of wealth, — ship-building is an engrossing interest, and there is no fête that calls forth the community like the launching of a vessel.

And no wonder; for what is there belonging to this workaday world of ours that has such a never-failing fund of poetry and grace as a ship? A ship is a beauty and mystery wherever we see it: its white wings touch the region of the unknown and the imaginative; they seem to us full of the odors of quaint, strange, foreign shores, where life, we fondly dream, moves in brighter currents than the muddy, tranquil tides of every day.

Who that sees one bound outward, with her white breasts swelling and heaving, as if with a reaching expectancy, does not feel his heart swell with a longing impulse to go with her to the far-off shores? Even at dingy, crowded wharves, amid the stir and tumult of great cities, the coming in of a ship is an event that can never lose its interest. But on these romantic shores of Maine, where all is so wild and still, and the blue sea lies embraced in the arms of a dark, solitary forest, the sudden incoming of a ship from a distant voyage is a sort of romance.

. . . The very life and spirit of strange, romantic lands come with her; suggestions of sandal-
wood and spice breathe through the pine woods;
she is an Oriental queen, with hands full of mystical gifts; "all her garments smell of myrrh
and cassia, out of the ivory palaces, whereby they
have made her glad." No wonder men have
loved ships like birds, and that there have been
found brave, rough hearts that in fatal wrecks
chose rather to go down with their ocean love
than to leave her in the last throes of her death-
agony.

A ship-building, a ship-sailing community has
an unconscious poetry ever underlying its existence. Exotic ideas from foreign lands relieve
the trite monotony of life; the ship-owner lives
in communion with the whole world, and is less
likely to fall into the petty commonplaces that
infest the routine of inland life.

Repression. There is a class of people in New England who betray the uprising of the softer feelings of our nature only by an increase of outward asperity — a sort of bashfulness and shyness
leaves them no power of expression for these
unwonted guests of the heart — they hurry them
into inner chambers and slam the doors upon
them, as if they were vexed at their appearance.

The Sab-
bath.
A vague, dream-like sense of rest and
Sabbath stillness seemed to brood in the
air. The very spruce-trees seemed to know that

it was Sunday, and to point solemnly upward with their dusky fingers, and the small tide-waves that chased each other up on the shelly beach, or broke against projecting rocks, seemed to do it with a chastened decorum, as each blue-haired wave whispered to his brother, "Be still — be still." . . .

Not merely as a burdensome restraint, or a weary endurance came the shadow of that Puritan Sabbath. It brought with it all the sweetness that belongs to rest, all the sacredness that hallows home, all the memories of patient thrift, of sober order, of chastened yet intense family feeling, of calmness, of purity, and self-respecting dignity, which distinguished the Puritan household. It seemed a solemn pause in all the sights and sounds of earth.

Early New England society. The state of society in some of the districts of Maine, in these days, much resembled, in its spirit, that which Moses labored to produce in ruder ages. It was entirely democratic, simple, grave, hearty, and sincere, — solemn and religious in its daily tone, and yet, as to all material good, full of wholesome thrift and prosperity. Perhaps taking the average mass of the people, a more healthful and desirable state of society never existed. Its better specimens had a simple, Doric grandeur, unsurpassed in any age.

THE MAYFLOWER.

A typical New England village. Did you ever see the little village of Newbury, in New England? I dare say you never did; for it was just one of those out-of-the-way places where nobody ever came unless they came on purpose: a green little hollow, wedged like a bird's nest between half a dozen high hills, that kept off the wind and kept out foreigners; so that the little place was as straitly *sui generis* as if there were not another in the world. The inhabitants were all of that respectable old steadfast family who made it a point to be born, bred, married, die, and be buried, all in the self-same spot. There were just so many houses, and just so many people lived in them; nobody ever seemed to be sick, or to die either, at least while I was there. The natives grew old till they could not grow any older, and then they stood still, and *lasted*, from generation to generation. There was, too, an unchangeability about all the externals of Newbury. Here was a red house, and there a brown house, and across the way was a yellow house; and there was a straggling rail fence or a tribe of mullein stalks between. The minister lived here, and Squire Moses lived there, and Deacon Hart lived under the hill, and Messrs. Nadab and Abihu Peters lived by the cross-road, and the old " Widder Smith " lived by the meeting-house, and Ebenezer Camp kept a shoemaker's shop on one side, and Patience Mosely kept a milliner's shop in front; and there was old Comfort Scran, who kept store for the whole town, and sold axe-heads, brass thimbles, licorice ball, fancy handkerchiefs, and everything else you can think of. Here, too, was the general post-office, where you might see letters marvelously folded, directed wrong side upwards, stamped with a thimble, and super-scribed to some of the Dollys, or Pollys, or Peters, or Moseses aforenamed or not named.

For the rest, as to manners, morals, arts, and sciences, the people in Newbury always went to their parties at three o'clock in the afternoon, and came home before dark; always stopped all work the minute the sun was down on Saturday night; always went to meeting on Sunday; had a schoolhouse with all the ordinary inconveniences; were in neighborly charity with one another, read their Bibles, feared their God, and were content with such things as they had, — the best philosophy after all.

The farm-house. Everything in Uncle Abel's house was in the same time, place, manner, and form, from year's end to year's end. There was old Master Bose, a dog after my uncle's own heart, who always walked as if he were studying the multiplication table. There was the old clock, forever ticking in the kitchen corner, with a picture on its face of the sun forever setting behind a perpendicular row of poplar-trees. There was the never-failing supply of red peppers and onions hanging over the chimney. There, too, were the yearly hollyhocks and morning-glories blooming about the windows. There was the "best room," with its sanded floor, the cupboard in one corner with its glass doors, the evergreen asparagus bushes in the chimney, and there was the stand with the Bible and almanac on it in another corner. There, too, was Aunt Betsey, who never looked any older, because she always looked as old as she could; who always dried her catnip and wormwood the last of September, and began to clean house the first of May. In short, this was the land of continuance. Old Time never took it into his head to practice either addition, or subtraction, or multiplication on its sum total.

UNCLE TOM'S CABIN.

Conscience in New England women. Nowhere is conscience so dominant and all-absorbing as with New England women. It is the granite formation which lies deepest, and rises out even to the tops of the highest mountains.

DRED.

Selling their disadvantages. "But these Yankees turn everything to account. If a man's field is covered with rock, he 'll find some way to sell it,

and make money out of it; and if they freeze
up all winter, they sell the ice, and make money
out of that. They just live by selling their disadvantages!"

POGANUC PEOPLE.

Yankee grit.   Zeph was a creature born to oppose, as
much as white bears are made to walk
on ice.

And how, we ask, would New England's rocky
soil and icy hills have been made mines of wealth
unless there had been human beings born to
oppose, delighting to combat and wrestle, and
with an unconquerable power of will?

Zeph had taken a thirteen acre lot, so rocky
that a sheep could scarce find a nibble there,
had dug out and blasted and carted the rocks,
wrought them into a circumambient fence, —
ploughed and planted and raised crop after crop
of good rye thereon. He did it with heat, with
zeal, with dogged determination; he did it all
the more because neighbors said he was a fool
for trying, and that he could never raise anything on that lot. There was a stern joy in
this hand-to-hand fight with Nature. He got his
bread as Samson did his honeycomb out of the
carcass of the slain lion. "Out of the eater
came forth meat, and out of the strong came
forth sweetness." Even the sharp March wind
did not annoy him. It was a controversial wind,
and that suited him; it was fighting him all the

way, and he enjoyed beating it. Such a human being has his place in the Creator's scheme.

Religious development. They greatly mistake the New England religious development who suppose that it was a mere culture of the head in dry, metaphysical doctrines. As in the rifts of the granite rocks grow flowers of wonderful beauty and delicacy, so in the secret recesses of Puritan life, by the fireside of the farmhouse, in the contemplative silence of austere care and labor, grew up religious experiences that brought a heavenly brightness down into the poverty of commonplace existence.

Family worship. The custom of family worship was one of the most rigid inculcations of the Puritan order of society, and came down from parent to child with the big family Bible, where the births, deaths, and marriages of the household stood recorded.

In Zeph's case, the custom seemed to be merely an inherited tradition, which had dwindled into a habit purely mechanical. Yet, who shall say?

Of a rugged race, educated in hardness, wringing his substance out of the very teeth and claws of reluctant nature, on a rocky and barren soil, and under a harsh, forbidding sky, who but the All-seeing could judge him? In that hard soul, there may have been, thus uncouthly expressed, a loyalty to Something Higher, however dimly perceived. It was acknowledging that even he had his master. One thing is certain, the custom of family prayers, such as it was, was a great comfort to the meek saint by his side, to whom any form of prayer, any pause from earthly care, and looking up to a Heavenly Power, was a blessed rest. In that daily toil, often beyond her strength, when she never received a word of sympathy or praise, it was a comfort all day to her to have had a chapter in the Bible and a prayer in the morning. Even though the chapter were one that she could not by any possibility understand a word of, yet it put her in mind of things in that same dear book that she did understand, — things that gave her strength to live and hope to die by, — and it was enough! Her faith in the Invisible Friend was so strong that she needed but to touch the hem of His garment. Even a table of genealogies out of *His* book was a sacred charm, an amulet of peace.

The kitchen fireplace. The fire that illuminated the great kitchen of the farmhouse was a splendid sight to behold. It is, alas, with us, only a vision and memory of the past; for who, in our days, can afford to keep up the great fireplace, where the backlogs were cut from the giants of the forest, and the forestick was as much as a modern man could lift? And then the glowing fire-palace built thereon! That architectural pile of split and seasoned wood, over which the flames leaped and danced and crackled like rejoicing genii — what a glory it was! The hearty, bright, warm hearth, in those days, stood instead of fine furniture and handsome pictures. The plainest room becomes beautiful and attractive by firelight, and when men think of a country and home to be fought for and defended, they think of the fireside.

The curfew. Though not exactly backed by the arbitrary power which enforced the celebrated curfew, yet the nine o'clock bell was one of the authoritative institutions of New England; and, at its sound, all obediently set their faces homeward, to rake up housefires, put out candles, and say their prayers before going to rest.

MY WIFE AND I.

Faculty. What Yankee matrons are pleased to denominate faculty, which is, being interpreted, *a genius for home life.*

CHAPTER IX.

MISCELLANEOUS.

OLDTOWN FOLKS.

The company room. It takes some hours to get a room warm where a family never sits, and which therefore has not in its walls one particle of the genial vitality which comes from the indwelling of human beings.

The turn of the tide. When you get into a tight place, and everything goes against you, till it seems as if you could not hold on a minute longer, never give up then, for that 's just the place and time that the tide 'll turn.

SAM LAWSON'S STORIES.

Little things. "Some seem to think the Lord don't look out only for gret things; but, ye see, little things is kind o' hinges that gret ones turns on. They say, take care o' pennies, an' dollars 'll take care o' themselves. It 's jest so in everything; and ef the Lord don't look arter little things, He ain't so gret as they say, any way."

LITTLE FOXES.

Sincerity and courtesy. Truth before all things; sincerity before all things; pure, clean, diamond-bright sincerity is of more value than the gold of Ophir; the foundation of all love must rest here. How those people do who live in the nearest and dearest intimacy with friends who, they believe, will lie to them for any purpose, even the most refined and delicate, is a mystery to me. If I once know that my wife or my friend will tell me only what they think will be agreeable to me, then I am at once lost, my way is a pathless quicksand. But all this being premised, I still say that we Anglo-Saxons might improve our domestic life, if we would graft upon the strong stock of its homely sincerity the courteous grace of the French character.

Flattery. Flattery is *insincere* praise, given from interested motives, not the sincere utterance to a friend of what we deem good and lovely in him.

HOUSE AND HOME PAPERS.

Household fairies. In fact, nobody wanted to stay in our parlor now. It was a cold, correct, accomplished fact; the household fairies had left it, — and when the fairies leave a room, nobody ever feels at home in it. No pictures, curtains, no wealth of mirrors, no elegance of lounges, can in the least make up for their absence. They are a capricious little set; there are rooms where they will *not* stay, and rooms where they *will*, but no one can ever have a good time without them.

SUNNY MEMORIES OF FOREIGN LANDS.

Cathe-
drals.   Cathedrals do not seem to me to have been built. They seem, rather, stupendous growths of nature, like crystals, or cliffs of basalt.

French
conversa-
tion.   Conversation of French circles seems to me like gambols of a thistledown, or the rainbow changes in soap-bubbles. One laughs with tears in one's eyes. One moment confounded with the absolute childhood of the simplicity, in the next one is a little afraid of the keen edge of the shrewdness.

The Ger-
mans.   These Germans seem an odd race, a mixture of clay and spirit — what with their beer-drinking and smoking, and their slow, stolid ways, you would think them perfectly earthy; but ethereal fire is all the while working in them, and bursting out in most unexpected little jets of poetry and sentiment, like blossoms on a cactus.

MY WIFE AND I.

Physiognomy of a house. Houses have their physiognomy as much as persons. There are commonplace houses, suggestive houses, attractive houses, mysterious houses, and fascinating houses, just as there are all these classes of persons. There are houses whose windows seem to yawn idly — to stare vacantly; there are houses whose windows glower weirdly, and glance at you askance ; there are houses, again, whose very doors and windows seem wide open with frank cordiality, which seem to stretch their arms to embrace you, and woo you kindly to come and possess them. . . .

Is not this a species of high art, by which a house, in itself cold and barren, becomes in every part warm and inviting, glowing with suggestion, alive with human tastes and personalities ? Wall-paper, paint, furniture, pictures, in the hands of the home artist, are like the tubes of paint, out of which arises, as by inspiration, a picture. It is the *woman* who combines them into the wonderful creation which we call a home.

When I came home from my office, night after night, and was led in triumph by Eva to view the result of her achievements, I confess I began to remember, with approbation, the old Greek mythology, and no longer to wonder that divine honors had been paid to household goddesses.

LETTERS.

Heaven. You *do* like to do good, and live a life worth living, and when you get to heaven you will always want to do exactly the thing by which you can best please the dear Lord. The fashions there in Heaven are set by Him who made himself of no reputation, and came and spent years among poor, ignorant, stupid, wicked people, that He might bring them up to himself, — and I dare say the saints are burning with zeal to be sent on such messages to our world, — I don't think they " sit on every heavenly hill," paying compliments to each other, but they are flying hither and thither on messages of mercy to the dark, the desolate, the sorrowful. That 's the way you 'll be when you get there, and spite of all you say about yourself, you 'll get to liking that sort of work more and more here.

His own house. I shall be glad when he is in a house of his own, — a man is n't half a man till he is.

FINIS.

INDEX OF BOOKS.

INDEX OF TITLES.

TEXT AND VERSE

FOR EVERY DAY IN THE YEAR.

Scripture Passages and Parallel Seléctions from the
Writings of JOHN GREENLEAF WHITTIER, arranged
by GERTRUDE W. CARTLAND. Tastefully bound in
cloth. 32mo, 75 cents.

This little book of helpful passages for every day is com-
posed of a verse from the Bible and a stanza of similar
character from Mr. Whittier's works arranged together.
The spirit of Mr. Whittier's poetry renders this association altogether harmonious, and the little book furnishes
a series of stimulating and consoling thoughts for the
whole course of the year.

Calculated to give strength and comfort to a wide circle, and to increase the poet's admirers.— *The Nation* (New York).

Very pretty, very cleverly done, and a success of which the au-
thoress may be proud. — *The Churchman* (New York).

FIVE EXQUISITE BIRTHDAY BOOKS

From the Writings of EMERSON, HOLMES, LONGFELLOW,
LOWELL, and WHITTIER.

These Birthday Books are of a high order of excellence. The most striking and beautiful passages in the
poems and prose works of these five illustrious writers
are arranged on the left-hand pages. On the right-hand
pages are given the names of distinguished persons born
and the record of memorable events occurring on the
same day. A blank space is left under each date. Each
of the books has a portrait of the author and twelve illustrations.

The price of the Birthday Books in cloth is $1.00
each, in flexible sealskin, $2.50.

*** For sale by all Booksellers. Sent by mail, post-paid,
on receipt of price by the Publishers,*

HOUGHTON, MIFFLIN AND COMPANY,
BOSTON AND NEW YORK.

Beckonings for Every Day.

A Calendar of Thought. Arranged by LUCY LARCOM, editor of "Breathings of the Better Life," etc. 16mo, $1.00.

It aims to give some of the most awakening and inspiring words of the great and good of all ages, and contains quotations from Browning and Phillips Brooks, from Robertson and Channing and Fénelon. The writer is catholic enough to include selections from James Martineau and Newman, and has been especially happy in finding new quotations both in prose and verse for her little volume. — *Boston Transcript.*

A book of devotional thoughts for daily use, of a much higher order than common. — *New York Evening Post.*

Flowers and Fruit

From the writings of HARRIET BEECHER STOWE. 16mo, $1.00.

This little book has been prepared with much intelligence and care, and includes a large number of striking and quotable passages from Mrs. Stowe's various works. They form a book of quite remarkable variety, admirable for brief leisure moments. Side titles are inserted in the text for each quotation, so that one can see at a glance what is the subject of each of the selections.

After Noontide.

Selected and edited by MARGARET E. WHITE. 16mo, $1.00.

A book containing passages designed to add sunshine and comfort to the afternoon of life. The selections are from numerous writers, of widely differing schools of literature and religion, but are rendered harmonious by the spirit of cheerful acceptance of the peculiar blessings which advancing years bring, and the recognition of the fact that though some burdens and infirmities may accompany these years, yet they are, or at least may be, full of cheer, of pleasant memories, and of serene hope.

** *For sale by all Booksellers. Sent by mail, postpaid, on receipt of price by the Publishers,*

HOUGHTON, MIFFLIN AND COMPANY,

BOSTON AND NEW YORK.

www.ingramcontent.com/pod-product-compliance
Lightning Source LLC
Chambersburg PA
CBHW021707210326
41599CB00013B/1549